Contents

T0286721

Executive Summary

Cities throughout the world have been at the forefront of a growing movement to implement "green" initiatives. Recognizing that urban centers spew most of the planet's greenhouse gas (GHG) emissions, proponents of sustainable communities press for resource conservation, clean air and water, and low-carbon energy and land use practices—with the entwined goals of averting climate disaster, righting longstanding environmental injustices, and creating more safe, healthy, equitable, and prosperous places to call home.

Activists gathered in Philadelphia, Pennsylvania, in August 2022 to call for a reduction in carbon emissions by Philadelphia Gas Works and a plan to incorporate climate equity with business development in this large legacy city. *Source: Emma Lee/WHYY.*

Only a handful of former industrial "legacy" cities in the United States have successfully made sustainability a policy priority and even fewer of them have been small (with populations of 30,000 to 75,000 people) or midsize (fewer than 300,000). Both small and midsize legacy cities grew rapidly in the nineteenth and early twentieth centuries before losing much of their industry and population in the decades after World War II. Their subsequent minimal engagement in sustainability cannot be attributed to lack of concern; rather, by definition, smaller legacy cities suffer from severe capacity constraints that limit their fiscal resources to the most urgent obligations, such as public safety, economic and workforce development, water delivery, and public retirement pensions.

While some large legacy cities have attracted philanthropic and private sector support for climate initiatives and have integrated sustainability into official municipal policies, smaller legacy cities have had proportionally fewer such resources available to them.

After decades of disinvestment and population loss, these cities also are plagued by overlapping difficulties that eclipse those of most other municipalities: vast swaths of stranded properties, crumbling infrastructure, widespread entrenched poverty, and pervasive environmental health issues in predominantly nonwhite neighborhoods. Any current or future sustainability initiatives in smaller legacy cities must address these fundamental challenges. Yet most legacy cities also have a wealth of real advantages that can be leveraged for green initiatives: access to fresh waterways and fertile farmland, dense urban fabrics with green space and potential for expanding mass transit, vacant lots that can be repurposed, and more.

While some large legacy cities have attracted philanthropic and private sector support for climate initiatives and have integrated sustainability into official municipal policies, smaller legacy cities have had proportionally fewer such resources available to them. Indeed, a 2017 Greater Ohio Policy Center report found that smaller legacy cities generally showed less evidence of post–Great Recession economic recovery than larger ones—suggesting a grim future for their tax base and the fiscal strength necessary to support robust low-carbon programs (Hollingsworth and Goebel 2017).

Yet "greening" smaller legacy cities is essential, despite these hurdles, for two primary reasons. First, these practices offer a promising place-based pathway for equitable economic and environmental rebirth, or "green regeneration." Second, although the argument is rarely made or is muffled by large, more successful cities in urban discourse—including that related to climate change—the United States needs places with agricultural and industrial production to drive its low-carbon future.

Smaller legacy cities help close the urban-rural divide by serving as regional economic centers and county seats. Their sense of scale differs from that of big-money, big-innovation global cities and thus offers a more modest antidote to our current lopsided winner-take-all economic terrain. Above all, their productive know-how and access to resources in nearby fertile farmland, forests, and water assets will be crucial to constructing a more sustainable, equitable low-carbon world.

In other words, greening is not just meant to assist cities that have fallen on hard times, or even to expand climate policy to all corners of the urban universe, as important as these goals are. Rather, in this era of rapid climate change, the United States needs its smaller legacy cities as much as—if not more than— smaller legacy cities need goods provided by other, more prosperous parts of the country.

For local officials and their partners across the government, nonprofit, business, and philanthropic sectors, this report offers actionable guidance for launching meaningful climate resilience, economic and equity programming, and for scaling early efforts effectively. Chapter 2 updates readers on the fast-changing world of climate policy and planning through three principal policy areas: climate resilience, environmental justice and racial equity, and green economic development. Chapter 3 explores how local officials and partners can use existing municipal policy levers to integrate all three policy areas by reforming land use practices, planning for blue-green infrastructure (BGI), developing brownfields and green buildings, and preparing for low-carbon energy build-out.

Recognizing the sheer magnitude of forthcoming federal investments in equitable climate infrastructure nationwide, and that smaller legacy cities have not received such transformative investment for at least three generations, chapter 4 proposes steps for launching public programming in partnership with the community. Finally, chapter 5 offers strategies and resources that link local sustainability efforts to regional and state partnerships, nonprofit intermediaries, and networks, and advises cities on how to attract and receive federal, state, and philanthropic climate funding. Along the way, these efforts may also lead to private sector investment that aligns with equitable climate goals.

Recommendations for greening smaller legacy cities, acting individually and in partnerships, are summarized in chapter 6. These ideas may also interest other poorly resourced or capacity-strained cities, suburbs, and towns across the country, whatever their size or history. Indeed, a strong regional, networked approach that centers climate resilience, racial equity, and green economic development can be useful to any region in mounting effective green regeneration policies, plans, and programs.

CHAPTER 1
Introduction

Smaller legacy cities like Trenton, New Jersey, can exert great regional influence, thanks in part to a distinct blend of existing assets like waterways and industry. *Source: Wirestock/iStock/Getty Images Plus.*

Smaller legacy cities have a complex conceptual history, in part because specialists in academia and philanthropy, policymakers, and the media have long disregarded them as a distinctive urban type. In a broad sense, these places were bypassed in favor of other urban policy preoccupations, from suburban sprawl to the role of global markets and tech innovation in local economic growth. Smaller legacy cities came into stronger policy focus with the 2007 collapse of the housing market and the ensuing Great Recession, which hit poor, depopulated legacy communities especially hard.

Reckoning with metropolitan, regional, racial, and economic disparities has generated an upsurge in cross-disciplinary scholarship on smaller legacy cities and urban policy think-tank initiatives (Connolly, Faulk, and Wornell 2022; MassINC n.d.; Mallach 2018a; Tumber 2012). Even influential academic opponents of place-based remedies for the woes of Rust Belt legacy cities now argue for regional wage subsidies and other local policy supports (Glaeser 2018). Big media outlets began paying closer attention too, but not until Donald Trump won the presidency in 2016, in part on a rhetorically populist platform opposing elites who outsourced US manufacturing to China and other developing nations.

The Position of Smaller Legacy Cities in a Low-Carbon Future

Greening and sustainability strategies can help smaller legacy cities restore and build upon existing assets, both natural and in the built environment, boosting them in the emerging climate-focused economy, as most are situated on fresh waterways and amid fertile farmland. Established before the automobile, legacy cities, mapped in figure 1.1, also tend to have "good bones" for sustainable cityscapes, with well-designed urban parklands and interconnected, walkable neighborhoods that can support large numbers of people living in dense housing. These cities are also home to architecturally substantial former industrial properties, or "brownfields," and other contaminated physical assets that can be remediated and sensibly redeveloped in equitable, low-carbon, climate-adaptive ways.

Smaller legacy cities also have distinct regional and institutional assets. With their long-established industrial firms and professional services, universities and hospitals, arts institutions, four-year and community colleges, courts, and county and even state government central offices (such as the state

The term "smaller legacy cities" refers to both small and midsize US cities that grew rapidly as productive centers of manufacturing, resource extraction, and food production in the nineteenth and early twentieth centuries. After World War II, these cities declined dramatically, losing their manufacturing infrastructure, substantial portions of their populations, and tax bases. Today, small legacy cities range in size, with populations between 30,000 and 75,000; midsize legacy cities have populations between 75,000 and 300,000 (Lincoln Institute of Land Policy 2022).

Researchers typically assume that legacy cities' median household incomes and educational attainment rates are lower than state averages, and that these cities' populations have dropped significantly and steadily from their peak. One-quarter of smaller legacy cities sit in a much larger, more successful city's "agglomeration shadow," and such proximity imparts characteristics unshared by stand-alone regional centers, so considerations for greening and other development efforts differ; this report thus focuses on stand-alone legacy cities.

Figure 1.1

Legacy Cities in the United States

capitols in Trenton, New Jersey, and Albany, New York), smaller legacy cities exert great influence on their regional communities. Even amid diminished conditions, they still play a crucial role in regional and state economies. In Ohio, for example, almost one-third of the state's population lives in and around such cities, generating 27 percent of the state's economic output (Greater Ohio Policy Center 2022; Hollingsworth 2016, output figure updated 2023).

Yet smaller legacy cities also face unique challenges. For example, while many larger legacy cities have had major job losses, they often retain Fortune 500 companies' headquarters or major research universities and medical centers—significant anchor employers and long-term community development partners that offer some reliable economic stability. Most smaller legacy cities, on the other hand, lack such assets and moreover have limited municipal capacity to implement new programs. In most, just one or two employees are available to coordinate sustainability initiatives, and they often have other job responsibilities. These private and public sector capacity constraints mean smaller legacy cities acting alone have difficulty preparing for low-carbon development and green regeneration.

After decades on the losing side of the global market economy, however, smaller legacy cities now have options for equitable climate action that play to their strengths. Survey data show that most Americans view climate change (71 percent), poverty reduction (88 percent), and energy system improvements (91 percent) as either important or top policy concerns (Pew Research Center 2023). Clean energy technology is coming into its own, with prices competitive with those of fossil fuels, greater efficiencies, and increasing battery storage capacity. Supply-chain disruptions amid the COVID-19 pandemic have also clarified the need to restore US manufacturing capability for critical national industries (Schwartz 2022). Furthermore, the racial reckoning in the United

States following George Floyd's murder in 2020 has sensitized more Americans to calls for racial and environmental justice.

The federal policy environment is also shifting dramatically and rapidly. For instance, in its first week, the Biden Administration launched the Justice40 Initiative via Executive Order 14008, which requires federal agencies to ensure that eligible disadvantaged communities receive 40 percent of the benefits from designated federal resources. Despite political headwinds, almost half of the $1.2 trillion Infrastructure Investment and Jobs Act (IIJA) of 2021 funded new transportation, water, and broadband infrastructure for the first time in a generation. As catastrophic fires, drought, and flooding later whipped through the country in 2022, the Inflation Reduction Act (IRA) launched more direct climate intervention by allocating approximately $390 billion over 10 years to environmental justice, domestic manufacturing, conservation programs, and clean energy projects (Dennis 2022). Finally, the 2022 CHIPS and Science Act invested $280 billion in US semiconductor manufacturing, research and development, and workforce education, among other activities that bolster the economy and national security (Kannan and Feldgoise 2022).

State and local governments and community-based nonprofits can apply for these and other grants and tax credits to advance local climate work such as greener building codes, urban forestry, green energy infrastructure, drought mitigation, energy efficiency, EV charging stations, and other initiatives to reduce greenhouse gas (GHG) emissions and toxic pollution, especially in communities of color. As the federal government remains committed to combatting climate change, environmental injustice, and domestic industrial decline, state and local governments must design a new generation of policies, partnerships, and resources that help their smaller legacy cities attract this new funding and investment.

Community gardens and urban farms can help legacy cities like Baltimore, Maryland, become more sustainable while also greening vacant and abandoned properties. *Source: JM Schilling.*

What Is Greening?

A term with deep roots in the modern environmental movement that gained legal traction in the early 1970s, "greening" suggests the need to protect natural systems from "gray" pollution caused by urban growth and heavy industry (Brinkley 2022). The growing awareness of climate change, the widening inequities of the global market economy, and the "urbanist" revival since the 1990s have led to a far more broad and complex understanding of greening in the context of sustainability and climate action.

Today, greening encompasses the following:

- **Urban sustainability** holds that resources in both nature and the built environment must be protected and replenished for future generations. It comprises three interdependent dimensions— ecological, economic, and social—without any one of which sustainable systems fall apart (Purvis et al. 2019). The social dimension necessarily emphasizes climate justice and equity, due to the disproportionate environmental and economic tolls on urban communities of color exacted by the segregationist legal structure that underwrote the era of suburbanization and deindustrialization following World War II (Rothstein 2017).

- **Decarbonization** mitigates the use of carbon and other GHGs by replacing fossil fuels with cleaner energy sources; extracting carbon from the atmosphere; eliminating most carbon-based products, services, or activities; and using energy more efficiently.

- **Low-carbon economic development** pursues policies and programs that incentivize products, processes, and services that are decarbonized, resource efficient, protective of biodiversity, and socially inclusive.

- **Blue-green infrastructure (BGI)** preserves and restores functioning ecosystems to ensure current and future access to essential water- and soil-based resources and to aid in adapting to climate change.

- **Smart Growth and New Urbanism** are distinct but related urban planning movements that call for land use and design principles that shape compact, dense, mixed-use, multimodal transit-oriented development; protect farmland and open space from suburban sprawl; and facilitate infill development, which is especially important in smaller legacy cities where many properties have been razed.

- **Equitable development** ensures that historically underserved communities and vulnerable populations will benefit from green investments in land use and in the built environment by fostering local resident engagement, leadership, and empowerment. It advances environmental justice remedies and protections in Black and Brown neighborhoods historically exposed to toxic contaminants.

- **Adaptive and whole-community resilience practices** integrate traditional resilience principles with community planning to help ensure sustainability investments are equitably distributed. This framework can further assist leaders when they engage underrepresented stakeholders in enhancing community assets and civic infrastructure to ensure they can withstand social, economic, and environmental shocks (Cowell 2013). Drawn from the field of disaster emergency management, whole-community resilience includes both short-term responses and long-term climate and economic adaptation planning for community assets—built, natural, and social—as well as for risks and vulnerabilities (Freitag et al. 2014; New Jersey Department of Environmental Protection 2022).

Timing or policy context might dictate prioritizing specific ecological, economic, or social elements of these various approaches within the scope of greening. In effect, urban sustainability is the overall object of greening, as communities pursue climate resilience and equitable economic regeneration.

A low-carbon future is necessary to avoid catastrophic climate change. If that reality is embraced and mitigation strategies are created and equitably developed, greening can foster the regeneration that has long eluded smaller legacy cities. That said, local officials need working knowledge of three key policy areas—climate resilience, environmental justice and equity, and green economic development. Their state and federal counterparts, as well as engaged nonprofit, business, and community partners, must also help advance the case for each, reframe their relationships with one another, and understand how conditions in smaller legacy cities can shape the implementation of greening policies.

A smaller legacy city, Harrisburg, Pennsylvania, has lost population but encourages density by rehabilitating aging housing and preserving its dense, walkable neighborhoods. *Source: peeterv/iStock/Getty Images Plus.*

CHAPTER 2
Integrating Greening Policy Areas

Climate change caused by our reliance on fossil fuels is bearing down on the planet with ever-increasing intensity, after decades of government inaction (see figure 2.1) have ensured it will continue unabated. The 2021 report from the United Nations Intergovernmental Panel on Climate Change (IPCC), based on the research of 234 interdisciplinary scientists, declared that the global community has only until 2030 to slow GHG emissions to levels that would enable the planet's climate to stabilize (Dennis and Kaplan 2021; Root 2021). In March 2023, the IPCC delivered an updated report with a "final" admonition for world leaders to act (Harvey 2023).

A popular clean power option in Asia, floating solar panel farms—like the one in Sayreville, New Jersey—can generate energy without a land footprint while also conserving water by preventing evaporation. *Source: Seth Wenig/ AP Photo.*

Figure 2.1

United States Billion-Dollar Disaster Events, 1980–2022 (Consumer Price Index Adjusted)

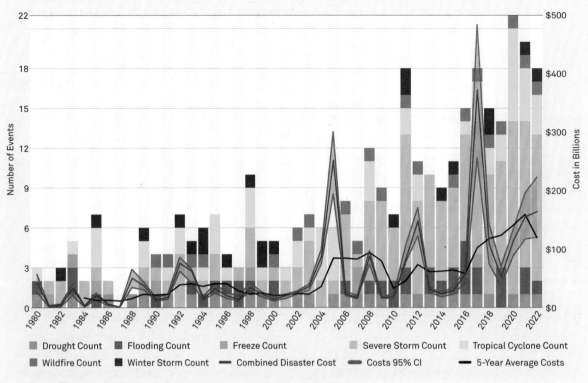

Source: Data from NOAA National Centers for Environmental Information (2023).

Fortunately, a cohesive portfolio of effective, efficient, and equitable sustainability initiatives could interrupt this cycle and build a greener future for cities of all sizes, economies, and geographies in the United States and elsewhere. In fact, global and national policymakers are counting on cities to deliver more than 50 percent of the necessary global reduction in GHGs (Coalition for Urban Transitions n.d.).

Meanwhile, communities across the country confront the interdependent challenges of climate resilience and equity due to increasingly vocal, organized, and documented racist incidents—and community reactions to them—even as they begin to redress economic, housing, and health disparities caused by structurally racist systems. Local officials must also grapple with

worrisome economic trends brought on by the COVID-19 pandemic, such as inflation, labor shortages, and supply chain disruptions. Smaller legacy cities feel these economic shocks more acutely than larger ones, given their history of deindustrialization and ongoing jolts from the 2007 mortgage foreclosure crisis (Mallach 2018a).

Pursuing green regeneration can provide smaller legacy cities with a more holistic framework to address these three concurrent challenges, though it will require unprecedented state and federal funding to build climate resilience, ensure environmental justice, and implement green economic development. This chapter explores the benefits and difficulties of integrating these three key greening policy areas and incorporating them into cities' policies and strategic plans.

The Complexities of Climate Resilience

Adapting to change is nothing new for smaller legacy cities. They have dealt with chronic challenges like population and job losses, vacant properties, racial injustice, and poverty associated with deindustrialization and suburban expansion for decades. Still, policy architects argue that the climate emergency requires a more strenuous form of adaptation called "climate resilience," whereby communities can prepare to withstand and rebound from different types of shocks while they implement mitigation technologies, programs, and strategies that curtail the use of fossil fuels—all while transitioning to low-carbon energy and economic infrastructure and more equitable planning practices.

Unlike the broader "climate action," the relatively new term "climate resilience" better reflects concepts relevant to local officials. As the Center for Climate and Energy Solutions explains, "Extreme weather events have shown that resilience is an essential component of any comprehensive climate action program because climate change is both a global and a hyperlocal issue. The causes and the broad impacts affect everyone on the planet, but resilience efforts must be executed at the asset, neighborhood, or individual level" (Center for Climate and Energy Solutions 2019).

Unlike the broader "climate action," the relatively new term "climate resilience" better reflects concepts relevant to local officials.

"Climate resilience" also encompasses the many overlapping impacts of climate change, now that its effects appear more rapid and expansive than previously predicted. This term not only covers simultaneous mitigation of GHG emissions and adaptation measures but also elevates protection of high-risk frontline communities (often comprising people of color) and vulnerable populations disproportionately affected by terrain changes (such as seniors, people with disabilities, and children). It further promotes sustainable economic development that ensures equitable access to jobs created by the transition to renewable energy. In other words, climate resilience is a pressing priority within greening policymaking.

A Note on Climate Migration

As the climate causes more severe weather, the United States and other countries already experiencing excessive heat, flooding, and sea-level rise will suffer more than others as climate change disrupts national and global supply chains, food and water supplies, and transportation systems. Experts predict that this ecological upheaval could drive climate refugees toward the Midwest and the Great Lakes region, where conditions may be more hospitable longer term but whose cities—long accustomed to population loss—are grossly unprepared for such an influx.

Nonprofit organizations like the Climigration Network have begun discussing the policy and planning actions that may be necessary for such communities to serve as relocation destinations following flooding, fires, drought, extreme temperatures, poor water quality and access, and other impacts of chronic natural disasters. Local officials should monitor climate in-migration in their own communities while following the growing literature and available resources on the subject ("Climigration Network" n.d.; Matthews and Potts 2018; Rajkovich et al. 2022; Steuteville 2016; Schneider 2021; Vock n.d.).

The Arc of Environmental Justice

Legacy cities' most environmentally hazardous neighborhoods face health threats from long-term exposure to lead paint and plumbing, exhaust fumes, toxic industrial pollutants and landfills, and poorly maintained housing. These are also overwhelmingly neighborhoods with a majority of Black and increasingly Latinx residents, since twentieth-century industry prompted large numbers of Black people from the American South to migrate north and into cities. The resulting demographic shifts entwined the lives of many Black Americans with the fate of the nation's urban manufacturing centers—including smaller ones. In these legacy cities, subsequent deindustrialization overlapped with discriminatory postwar housing and suburbanization policies, such as exclusionary zoning, redlining, urban renewal, and urban highway building, which generated many socioeconomic, environmental, and health inequities that prevail today. The word "legacy" covers this history, too.

Chief among these inequities are chronic health issues and high mortality rates (American Lung Association 2022). Though risks persist for all neighborhood residents, significant research has identified race as the strongest indicator for environmental health issues. Furthermore, children are especially vulnerable to the effects of toxic exposure, due to their developing physiology and close interactions with the environment (American Public Health Association 2019). Thus, at its foundation, the environmental justice movement pushes for individuals' basic human right to a clean, health-fostering environment (Agyeman et al. 2016; Bullard 1993).

The movement has learned hard lessons from its own history, earning mainstream recognition in the early 1990s. Spurred by a serious but unsuccessful 1982 grassroots battle led by the National Association for the Advancement of Colored People (NAACP) to prevent a toxic landfill siting in Warren County, North Carolina, a series of lawsuits tested the limits of new federal environmental laws, and local resistance emerged across the country (McGurty 2009). In 1991, more than 1,000 people from all 50 states and a few other countries attended the First National People of Color Environmental Leadership Summit, which rolled out the 17 Principles of Environmental Justice comprising the movement's founding document (First National People of Color Environmental Leadership Summit 1991).

Attendees of the First National People of Color Environmental Leadership Summit rallied at the U.S. Capitol in October 1991 on behalf of the summit's 17 adopted principles of environmental justice, which recognized Black, Brown, and Indigenous leadership in protecting the natural world. *Source: Dr. Robert Bullard.*

Higher land surface temperatures are common in historically redlined areas, meaning communities of color suffer disproportionately from a deficit of accessible green amenities—and that residents must resort to extreme measures to combat the intensity of urban heat. *Source: PeskyMonkey/ iStock Photo.*

Movement pressure and public opinion led to the 1992 establishment of the Office of Environmental Equity, later renamed the Office of Environmental Justice at the US Environmental Protection Agency (EPA). In 1994, President Bill Clinton issued Executive Order 12898, directing executive agencies to incorporate environmental justice into their missions (Agyeman et al. 2016; US EPA 1994). Still in effect, Clinton's order generally operates in conjunction with Title VI of the 1964 Civil Rights Act, which prohibits discrimination based on race, color, or national origin among federal funding recipients. More recently, in 2021, President Joe Biden's Justice40 Initiative pledged unprecedented federal investment in remedying environmental injustice. The Biden Administration subsequently established the White House Office of Environmental Justice in April 2023 to coordinate policies and ensure accountability across federal agencies (White House 2023a).

To date, however, federal levers and legal recourse have been largely insufficient, and it remains to be seen how effective recent policy changes will be. Consider the following legal and administrative weaknesses:

- A 2015 study of the EPA's Environmental Appeals Board found that the board had never blocked a permitting decision based on an environmental justice consideration (Gauna 2015).
- As of 2016, eight environmental justice lawsuits based on the Equal Protection Clause of the 14th Amendment had been filed, but all failed in part because of the Supreme Court's 2001 *Alexander v. Sandoval* decision, which ruled, "a government action that might have a discriminatory impact is not unconstitutional unless the decision-maker had a discriminatory *intent*, which is something that is very hard to prove" (Cole and Foster 2001, italics in original).
- Of the 298 Title VI complaints against perceived discrimination by a public agency using federal funds filed as of January 2014, only one was upheld.
- With the appointment of more conservative judges during the Trump Administration, many courts are further primed to reject claims of racial discrimination in environmental cases.

This record indicates that relying on federal courts and regulatory bodies is insufficient for spurring equitable green regeneration in smaller legacy cities. Although more than 30 states have adopted environmental justice initiatives (including product bans, environmental charters, targeted land use measures like buffer zones, enhanced public health guidance, and environmental review processes), they too often lack the standards, accountability, staffing, or legal authority to generate the necessary systems change. Though federal and state efforts (or lack thereof) can have significant impacts on environmental injustice, local accountability remains. Officials who control land use decisions are often heavily influenced by developers seeking cheap land, so they share much of the responsibility for past and future toxic exposure in Black and Brown communities.

Equitable development is a key, though not sole, remedy. A place-based approach to urban planning and community development—in contrast to policies

COVID-19 and Toxic Exposure in "Sacrifice Zones"

Kimberlé Williams Crenshaw, cofounder and executive director of the African American Policy Forum, astutely describes the "brutal predictability" of racial disparities in health and wealth as a "bald political calculus—one that ultimately pivots on an 'acceptable' number of deaths in poorer nonwhite communities" (Crenshaw 2020). In her essay "The Unmattering of Black Lives," Crenshaw further links COVID-19 lethality among Black Americans to environmental racism and healthcare disparities that long predate the pandemic.

The rise of industrialization brought hazardous waste and pollution to low-income areas, usually communities of color or tribal lands removed from more affluent white populations. These communities often become "sacrifice zones," areas permanently impacted by environmental degradation through unwanted land uses and pollution (Lerner 2012). Indeed, as journalist Naomi Klein observes, "running an economy on energy sources that release poisons as an unavoidable part of their extraction and refining has always required sacrifice zones—whole subsets of humanity categorized as less than fully human, which made their poisoning in the name of progress somehow acceptable" (Klein 2015).

that induce residents to relocate to stronger labor markets—this family of strategies asserts that, to facilitate regional prosperity, all communities must have access to employment, economic opportunities, social services, food, safe and affordable housing, education, transportation, and healthy built environments (Eley 2020). It is necessary to close racial disparities in these realms to ensure equitable access to the benefits of climate resilience and green economic development.

Equity demands critical analysis of government decision-making and a cohesive approach across multiple policy domains. Although toxic exposure remains a major longstanding grievance, equitable development looks beyond pollution toward how racial and socioeconomic inequities increase vulnerability to climate change impacts and limit participation in urban greening, green economic development, and green technologies.

Though some government-centric strategies may be necessary to mitigate immediate short-term harms, environmental justice advocates argue that other such strategies may inhibit structural change, noting that the state has played a central role in perpetuating environmental injustice (Karner et al. 2020). Local governments should be aware of these tensions and receptive to information gathering and appropriate power-sharing arrangements. In Providence, Rhode Island, for example, the city's sustainability department partnered with the Racial and Environmental Justice Committee of Providence to build trust and strengthen municipal relationships with frontline communities of color, advancing more collaborative decision-making processes (City of Providence 2019).

Members of Concerned Citizens of St. John Parish, in Reserve, Louisiana, demand justice by warning residents about the risk of toxic exposure from a nearby chemical plant. *Source: Julie Dermansky.*

Public-Private Tensions in Green Economic Development

Green regeneration in smaller legacy cities will only succeed if policies and programs lead to decent, accessible jobs for local communities. Yet smaller cities, in general and especially legacy cities, lack the private capital, public resources, or appropriate policy levers for green economic development (Mallach 2022). They must instead collaborate extensively with relevant federal, state, and regional governing agencies while retaining and attracting private sector investment and strategic support from nonprofits, universities, and philanthropy.

Complicating matters, three major misconceptions cloud the very idea of green economic development (GED), and local officials must clarify its meaning—for themselves and for their residents—to move forward. First of all, while GED may generate "green jobs," it is far more accurate to say that GED targets industries that safeguard environmental resources while generating further economic development and new jobs in both green and other sectors through the multiplier effect (Bohnenberger 2022; DiNapoli 2022; Fitzgerald 2010; National Association of Counties 2010).

Sea-level rise threatens many houses and businesses in coastal legacy cities such as Norfolk, Virginia, with its extensive waterways and oceanfront properties. *Source: Andyd/E+/ Getty Images.*

Construction, leisure, hospitality, and education, though not themselves green industries, often benefit from GED. For example, a large region with a favorable wind shed might attract public and private investment in several utility-scale wind farms that support nearby turbine manufacturers and suppliers, who make some parts destined for auto assembly, other construction projects, and sewer expansion. Tracking the explicitly "green" outcomes of this process can be knotty at best; even the US Bureau of Labor Statistics has had trouble defining and tracking green jobs (Bohnenberger 2022; US Bureau of Labor Statistics 2013).

GED also differs from traditional economic development growth models, which are principally concerned with expansion of exportable goods and output measured by cash value. While both concepts are tied to quantitative methods for establishing benchmarks and measuring change over time, economic

development additionally considers qualitative data such as surveys and interviews, and determines the strength of the local economy using many more indicators, including quality of education, the housing market, and employment compensation and turnover. Green economic development covers these indicators but also ties them to measurable system changes in infrastructure, consumption patterns, transportation use, toxic exposure, and other trends relevant to environmental impacts.

A third misconception concerns the role of government. Since the country's earliest days, governments of various scales have spearheaded big infrastructure projects, from the Erie Canal and the first transcontinental railroad network to the electric utility and interstate highway systems. These activities ran the moral gamut from providing electricity to the

The downtown renaissance in Erie, Pennsylvania, has relied on improving existing infrastructure and leveraging waterfront assets as foundations for its redevelopment plan and strong public-private partnerships. *Source: Corey Cook.*

impoverished rural American South during the New Deal era to forcibly removing Indigenous peoples from their land, resulting in a deeply mixed legacy. Indeed, these public projects also assumed the high-risk initial financing and policymaking that cleared the way for private sector build-out and profit making.

Since the 1980s, some conservative critics and political leaders have viewed the federal government and the free market as antagonists; today, public-private partnerships generally defer to the private sector, with government as funder and facilitator rather than guide. Yet private businesses are accountable only to their owners and shareholders and they thus create what economists call externalities, such as water pollution or unhealthy working conditions, that market corrections alone do not address. Climate change is one giant, global, time-sensitive externality; if markets could target the problem at the necessary scale, they would have done so already.

To act, the American private sector requires market signals on a scale that only the federal government can provide. Of course, this involves advisory coordination with business interests, which in turn necessitates anticorruption measures. Still, good local GED programming can bridge the extreme government/free market divide that has prevailed ideologically for more than 40 years.

Making traditional economic development strategies and tools more responsive to emerging green economic opportunities could center climate resilience and environmental justice in a future policy vision. For now, however, many local officials still rely on basic local economic development strategies and tools such as tax increment financing (TIF) and tax incentives. They also traditionally defer to regional chambers of commerce, commercial banks, workforce investment boards, and other business organizations to steer things, rather than looking into their own diverse communities for ideas and guidance.

Cities and counties can take several fundamental GED actions to shape public and private sector investments in constructing low-carbon infrastructure and green buildings, applying a restorative approach to systemic injustices, making better use of natural assets, and creating jobs—including "green" ones. Detailed in chapter 5, these steps take advantage of local networks, knowledge, and businesses while prioritizing equity considerations and intersectoral coordination.

The Journey to Sustainability

Green regeneration doesn't happen overnight, especially for smaller legacy cities, only a handful of which now have a cohesive portfolio of sustainability policies, plans, and programs (Schilling and Velasco 2020). Transitioning communities and regional economies to new, low-carbon energy systems while adapting to the early yet devastating effects of climate change is a massive endeavor. Equitable sustainability initiatives can, however, provide economic opportunities for all legacy city residents if local government officials can leverage place-based assets and on-the-ground engagement that expand community power and voice. Such transformative leadership requires cultivating community champions, facilitating policy coordination, expanding government capacities, and continually developing new knowledge and partnerships.

Transformative leadership requires cultivating community champions, facilitating policy coordination, expanding government capacities, and continually developing new knowledge and partnerships.

As they advance sustainability initiatives, city officials, citizen advocates, and community stakeholders must assess existing efforts (or lack thereof) so they can set, measure, and achieve short- and long-term goals. Research of successful legacy cities reveals a three-generation evolution of sustainability policies, plans, and programs that often starts with expanding basic environmental programs and services. In the second phase, local officials begin to address contemporary sustainability challenges by charting out specific plans, dedicating more staff and resources to greening programs, and establishing collaborations with nonprofit partners and community members across urban sustainability's three primary pillars.

Building on these early-generation policies and programs, local governments must ultimately blend climate resilience, environmental justice, and green economic development initiatives into a mutually reinforcing system that leads to equitable job growth, a stronger fiscal base, and planetary preservation. A handful of large and midsize legacy cities have entered this third-generation programming phase. (Schilling and Velasco 2020).

Table 2.1 outlines the local government policy landscape and subsequent capacities that expand and evolve over time, depending on the actors and priorities within a dynamic policy ecosystem. (See chapter 5 for additional discussion.)

A successful progression through these "generations" will enable policymakers in small legacy cities to align their managers, frontline staff, and partners with the goals and priorities of environmental justice advocates, civic and business leaders, and community-based organizations. Everyone involved must understand the challenges and opportunities that come from third-generation green regeneration—and commit to this collaborative journey (Schilling and Velasco 2020).

INSIGHTS FROM BENCHMARK LEGACY CITIES

For many legacy cities, a critical step is assessing where they stand and where they need to go on their path to green regeneration—and how to get there. Most legacy cities, even smaller ones, have

Table 2.1

Green Policy Evolution in Local Governments

Generation	First Generation	Second Generation	Third Generation
Policy Actions	• Green government operations • Redevelop brownfields and offer related job training • Build recycling and stormwater infrastructure • Enact smart growth planning and codes • Enable urban greening and vacant property reclamation • Extend green business networks and best practices	• Create Climate Action Plans (CAPs) and GHG inventories • Implement green building codes and standards • Establish green development zones and ecodistricts • Create area-wide brownfields redevelopment plans and multipurpose brownfields grants • Collaborate with regional, state, and national intermediaries • Partner with local universities and/or nonprofits	• Launch and enforce sustainability plans, sustainable development codes, climate resilience plans, climate equity plans, and environmental justice commissions, screens, and codes • Prioritize renewable energy, decarbonization, and energy efficiency • Expand policy ecosystem for green industries, businesses, and jobs • Elevate environmental justice in all green regeneration actions
Capacity-Building Actions	• Hire part-time sustainability coordinator or expand a current role to include such responsibilities	• Hire full-time sustainability manager with part-time staff • Offer peer learning through professional green networks and local government associations	• Hire full-time sustainability director • Dedicate office and full-time staff to sustainability • Expand capacity through regional, state, and national intermediaries

Twelve stories above street level, the blue-green roof of Philadelphia's Cira Green manages over 700,000 gallons of stormwater annually and includes an elevated park with more than 30,000 square feet of public green space. *Source: Brandywine Realty Trust.*

already taken a range of first- and second-generation approaches, such as adopting a climate action plan (CAP), redeveloping brownfields, or assigning sustainability programs to a part-time coordinator. Few, however, have cultivated the requisite leadership and management or assembled the technical capacities and resources that integrate the three primary policy areas—climate resilience, environmental justice, and green economic development. Indeed, only a few midsize or large legacy cities have effectively implemented a comprehensive suite of sustainability policies, plans, programs, projects, and practices (Schilling and Velasco 2020).

The early waves of green regeneration in legacy cities gained traction with the Obama Administration's multipronged responses to the economic distress of the Great Recession, such as the Strong Cities, Strong Communities initiative and the Hardest Hit Fund. In 2009, the Obama Administration also formed the Partnership for Sustainable Communities to improve interagency coordination of federal sustainability policies and programs, such as the Sustainable Communities initiative. Cleveland, Philadelphia, and Pittsburgh—all larger legacy cities—are good models for how to align strong political leadership, external grants, and the capacity and expertise of innovative nonprofit intermediaries to build out all three generations of their sustainability programming.

As early adopters, large and midsize legacy cities can offer insights into how evolving policy integration—and recent infusions of federal and state economic, energy, and climate resources—might play out in smaller legacy cities.

The Greening of Philadelphia, Pennsylvania

By the end of his administration in 2016, Mayor Michael Nutter had made Philadelphia into one of the greenest cities in the country by adopting a portfolio of sustainability initiatives, including creating the City's Office of Sustainability and its first sustainability plan, and the water department's award-winning green infrastructure plan. These green elements continue to inform the implementation and revisions of the city's comprehensive land use plan and zoning code.

The city also collaborates with nonprofits on sustainability priorities such as greening vacant lots with the Pennsylvania Horticultural Society, facilitating green community development, developing a citywide green workforce, and advancing a just and green business community. Building on this foundation, Philadelphia's third-generation activities include institutionalizing the Office of Sustainability and adopting its climate adaption plan (2015); developing its racial equity action plan and municipal energy master plan for the built environment (2017); launching the city's first racial equity strategy (2020); and releasing a climate action playbook (2021) and convening its new environmental justice advisory commission (2022) (Adaptation Clearinghouse 2015; Duchene 2022; Philadelphia Water Department n.d.).

CHAPTER 3

From Sustainability Policy to Local Implementation

The City of Providence, Rhode Island, launched a climate action plan that included measures to stop displacement and erasure of minority communities, including Narragansett tribal members like those depicted in "Still Here," a 2018 mural by street artist Gaia. *Source: Gaia.*

Local officials entrusted with governing smaller legacy cities have a lot on their plates. Most do not have sufficient resources or staff to develop robust sustainability initiatives beyond including individual greening programs with current functions, like adding recycling to waste removal programs, greening city operations, or executing court-mandated interventions such as lead abatement in drinking water or upgrading antiquated stormwater systems. In many states, especially throughout the Midwest and South, local officials may also lack support from their colleagues, the community, state governments, or the courts to exert strong leadership or forge new partnerships on climate action and equity.

Further, climate action demands many changes, such as upgrading the national electric grid, that are simply beyond local control. In these cases, local efforts by small nonglobal cities can seem futile, as can making the case to neighborhood communities, metropolitan and rural areas, state powers, private sector players, philanthropies, and nonprofits. Yet their support is critical if smaller legacy cities are to decarbonize fairly and speedily.

That said, local officials have many policy and planning levers they can pull to both mitigate and adapt to the effects of climate change, positioning their residents to weather the often-literal storms ahead and presenting a persuasive vision of practical community benefits. As smaller legacy city government leaders know all too well, however, they still need cross-sector partners, as well as strategic guidance, technical capacity, and resources.

The next two chapters chart out actions local officials can take on their own based on current practice, including land use—a key focus of municipal powers—as well as green infrastructure, business development, and other baseline measures. Chapter 5 then covers how partnerships can help local officials broaden their efforts, regionally and through city policy, plans, and programs that can equitably leverage recent federal funding for infrastructure and climate projects.

Local officials have many policy and planning levers they can pull to both mitigate and adapt to the effects of climate change, positioning their residents to weather the often-literal storms ahead and presenting a persuasive vision of practical community benefits.

Climate Resilience in Sustainability Plans and Codes

Comprehensive land use planning, development processes, and zoning codes provide the policy foundation for smaller legacy cities to address the intersections of climate resilience, environmental justice and equity, and green economic development. Most local governments have some authority over land use, with local land development codes and processes able to shape communities' responses to changing weather patterns and natural disasters, address racial disparities, and facilitate green economic development. For example, land use policies can restrict development in areas prone to wildfires or flooding and incentivize upgrades to gray "hard" infrastructure such as dams, wastewater treatment facilities, and roads, with more energy-efficient or nature-based features. They can also redress the racist legacies of past and present land use practices, ensuring that climate equity is built into greening strategies.

TYPES OF SUSTAINABILITY PLANS

Municipalities have several types of formal plans that can holistically feature sustainable elements, but they often resort to supplementary stand-alone green plans, creating a patchwork that addresses sustainability issues piecemeal, led by separate departments.

Climate action plans focus primarily on decarbonization measures and signal that a local government is serious about climate change, with GHG inventories a common first step toward greening city facilities. However, sustainability practitioners now question the efficacy and relevance of CAPs for smaller cities, given that most emission sources lie beyond their jurisdictions and regulatory reach (Armstrong et al. 2021).

Climate equity plans elevate race and equity as major determinants of climate resilience policy actions. A core principle of this relatively new plan type is that equitable processes drive equitable outcomes and thus expand the reach of sustainable practices. These plans often outline policy and program actions that cover housing and gentrification, access to clean energy and transportation, community health, climate resilience, and the local and regional economy (City of Providence 2019). Chapter 4 includes a discussion of how Providence, Rhode Island, put the community at the center of the planning process and set the stage for the resulting Climate Justice Plan.

Local resilience plans often begin with a systematic, community-wide risk assessment in the context of disaster planning. Since 2012's Hurricane Sandy, the distinction between climate and resilience planning has increased. As climate-induced disasters mount, resilience plans have broadened. Tornado-damaged Springfield, Massachusetts, for example, developed a combined climate action and resilience plan that included upgrading flood control systems and natural restoration in a large urban watershed. It also planned for better job training and business development in surrounding low-income neighborhoods while reducing GHG emissions 80 percent by 2050 (City of Springfield 2015; City of Springfield Office of Community Development 2017).

Many cities confront the challenge of how to align and integrate multiple plans that may each focus on one or two sustainability policy areas but with different priorities, goals, and actions. Multiple plans—and their potential to conflict—can confuse government officials and send the wrong message to the community, stymieing progress. A number of cities have solved this problem by integrating goals for sustainability, climate and resilience, and equity throughout their comprehensive land use plans, in essence creating an overarching, comprehensive sustainability plan (Rouse 2022).

Many comprehensive plans already contain effective smart growth strategies such as revising single-family zoning regulations, removing onerous parking requirements, establishing mixed-use zoning, requiring Low-Impact Development (LID) and energy-efficiency standards for new development, or adopting "complete streets" programs (ChangeLab Solutions n.d.). As more communities update these plans, several—including Springfield, Massachusetts, and Syracuse, New York—have adopted companion plans for climate action, resilience, or equity that more consistently address topics with land use implications, like food

Following a devastating 2011 tornado, the city of Springfield, Massachusetts, improved its flood control and watershed management to better gird against future disasters caused by climate change. *Source: Jessica Hill/Associated Press.*

Residents of Baltimore, Maryland, called for zoning changes at a 2018 rally demanding that the city ban crude oil transport through densely populated areas. *Source: Clean Water Action.*

and water security, natural hazards, green buildings, renewable energy infrastructure, and equitable development.

Several regional and state governments have also adopted their own versions of climate action, resilience, and sustainability plans that smaller legacy cities can leverage, as discussed in chapter 5. In planning specifically for smaller legacy cities, though, these practitioners should consider: Are these different plans consistent and supportive of one another? Can smaller legacy cities tailor these plans to address their special socioeconomic and environmental challenges? Are there networks or programs that can help build capacity to implement local land use plans and maintain community engagement with the process?

GREEN DEVELOPMENT CODES AND EQUITY ZONING

Over the past 30 years, the Smart Growth and New Urbanism movements catalyzed initiatives to reform mid-twentieth-century zoning designed for separate functional uses (typically commercial, industrial, and residential) that accommodated manufacturing, the automobile, and large-lot single-family housing. Their changes elevated model land use codes, such as unified development ordinances and form-based codes (US EPA 2013). These reforms made it easier to construct mixed-use infill development and low-carbon transportation (mass transit, walking, biking, wheelchairs, scooters), while also establishing design guidelines for urban greening and green building codes.

Zoning codes also specify densities, as with transit-oriented development (TOD), which incentivizes developers to build intensively near multimodal transportation hubs or commercial centers. Climate-conscious land use restrictions might also disincentivize or even prohibit certain land uses and structures that increase GHGs or other climate risks, such as allowing new homes in flood plains. Zoning, however, only sets baseline rules for land development within the confines of federal and state laws governing private property rights and the market decisions made by land developers, financial institutions, builders, businesses, and homeowners. Reforming an entire zoning code can also take years and requires significant staff time and resources—or outside consultants, a practice that can inhibit trust and effective community engagement. Within the context of smaller legacy cities' capacity constraints, these and other barriers become magnified.

The Sustainable Development Code's field-tested template offers a good starting point, with specific code sections that cover transportation and mobility, land use and community character, community health, natural hazards, and other topics. A nonprofit team of law professors and practitioners developed the model code and continue to provide technical assistance, alongside useful tools for reviewing existing zoning and development codes and for improving how socioeconomic, environmental, and discriminatory practices play out in the built environment (Sustainable Development Code n.d.).

In lieu of greening an entire zoning code, municipalities commonly rely on overlay districts, which can be dedicated to particular uses such as environmental protection, historical preservation, and manufacturing. They may also try various long-term pilot projects—including "ecodistricts" devoted to integrating energy,

water, waste, transportation, and land use systems on the neighborhood or district scale. Although an umbrella organization called EcoDistricts offers guidance, training, and formal certification, ecodistricts can assume a variety of forms (Argerious 2022; Ecodistricts.org n.d.).

Zoning for climate resilience must also ensure equitable treatment and benefits for communities of color that have survived decades of exclusionary zoning, segregation, urban renewal, and disinvestment policies. Even in cities with considerable resources, zoning for equitable environmental outcomes remains a challenge. The need to screen current zoning codes for inequitable practices is great, however. They are often layered with mid-twentieth-century segregationist ordinances and "downzoning" provisions that authorize industrial uses in neighborhoods of color.

The Tishman Environment and Design Center's 2019 review of 40 local zoning policies identified six policy classes of environmental justice–related zoning reforms and offers a template for designing and implementing local equity screens, comprising

Adopted in 2017, the "Green Code" is one of many form-based code changes the City of Buffalo, New York, has made to incentivize sustainability, affordability, density, and transportation options. *Source: SerrNovik/iStock/Getty Images Plus.*

- bans on specific types of polluting facilities typically sited in historically underinvested communities;

- broad environmental justice policies that incorporate relevant goals and considerations into a range of municipal activities;

- environmental review processes applied to new developments;

- equity measures and protections incorporated into future development via comprehensive plans, overlays, or green zones;

- targeted land use measures such as buffer zones that address existing sources of pollution; and

- enhanced public health codes surrounding both existing and new sources of pollution that impact public health (Tishman Environment and Design Center 2019; Mulvihill 2020).

Transformational zoning reforms require substantial political will, bold adherence, and regulatory teeth that ensure strict accountability over time. Plans for equitable development must deliberately and explicitly engage the community in order to create the conditions for material change.

As communities change their zoning and land development codes, local policymakers must also consider how land use can mitigate potential "green gentrification" caused by market forces. Most smaller real estate markets in legacy cities are weak enough that residents won't be put in immediate danger of displacement as a by-product of greening; such cities also tend to enthusiastically welcome investors. As equity investors speedily buy up property across the United States, however, local officials would be wise to explore tools such as income-based property tax relief, subsidized rental costs, and capped property taxes for longtime residents, as well as housing land trusts (Immergluck 2017).

Only a handful of third-generation green zoning codes currently exist for others to use as models. Buffalo,

New York, created its first comprehensive update since 1953 with the Unified Development Ordinance or "Green Code," which blends equity, sustainability,

---

The Challenges of Zoning Against Environmental Injustices

Baltimore, Maryland: Crude Oil Terminal Prohibition Ordinance

In 2014, residents of the environmentally degraded Curtis Bay neighborhood organized against Targa Terminals' proposed crude oil terminal in the Fairfield area of South Baltimore, prompting Targa Terminals to withdraw its application amid related legal challenges. The residents then urged the city to study the impacts of crude oil trains, such as the health impacts of exposure to volatile organic compounds (VOCs) and the risk of train explosion. Though their bill was rejected by the city council, the Brooklyn–Curtis Bay Coalition continued to organize, partnering with national groups like Clean Water Action. Inspired by similar West Coast measures, residents proposed local zoning codes to ban specific fossil fuel infrastructure—and won. In 2018, the City of Baltimore passed the Crude Oil Terminal Prohibition Ordinance, becoming the first East Coast city to use a zoning code to ban new crude oil terminals (Tishman Center 2019).

Cincinnati, Ohio: Enforcement Blockages

Toxin-emitting industries have remarkable lobbying and political power and are an intimidating adversary for many local governments and activists. The City of Cincinnati, Ohio, attempted in 2009 to implement an Environmental Justice Ordinance with police enforcement powers, following years of organizing from local groups. After a miserable fight with the Cincinnati USA Regional Chamber® (of commerce) and other business groups, however, the ordinance was delayed indefinitely just a year later. Today, what remains of the would-be "model" ordinance is only a public notification process (De Guire 2012; Knight 2018).

Students from the Miller Street School plant a rain garden as part of a Greater Newark Conservancy program to improve green infrastructure and create transitional jobs. *Source: Jeremiah Bergstrom with Rutgers Cooperative Extension.*

climate resilience, and New Urbanist principles. It is also deliberately democratic in form: jargon-free, heavy on accessible graphics, and one-fifth the size of its bulky predecessor. Adopted in 2017 after more than six years of public engagement, the Green Code promotes walkability, density, and mixed-use form-based design that centers sustainability and affordability (Sommer 2017). The code also mitigates displacement by attaching affordability requirements to density incentives to ensure access for all residents. Citywide form-based codes are coming into greater use in smaller legacy cities, with Hartford, Connecticut, and South Bend, Indiana, following Buffalo as recent winners of the prestigious Driehaus Award from Smart Growth America's Form-Based Codes Institute (Hope 2021).

BLUE-GREEN INFRASTRUCTURE: NATURE'S BREATHING ROOM

Cities are covered with miles of impervious sur-faces—buildings, sidewalks, parking lots, roads, and more—that absorb or generate heat and cause exces-sive stormwater runoff that pollutes rivers, lakes, and oceans. Beneath the surface are networks of aging "gray" infrastructure that deliver safe drinking water, manage floods, and divert harmful stormwater

pollution. As communities experience more extreme and prolonged climate-driven weather events, local governments can respond by investing in and aligning their extensive networks of blue (water) and green (natural or land-based) infrastructure, or BGI.

As a result of decades of neglect, lack of maintenance, and insufficient federal and state government funding, many older communities' water systems, including those in nearly all smaller legacy cities, desperately need upgrading to handle more intense storms, droughts, and flooding fueled by climate change (Vedachalam et al. 2020). For example, sustained heavy rainfall in Eastern Kentucky took the lives of 44 people in 2022 and inflicted nearly $1 billion in flood and storm damage (Morford 2023). Most smaller legacy cities in the Northeast and Midwest must also upgrade decaying drinking water systems and lead service plumbing, a particularly widespread deficiency in neighborhoods of color, epitomized by the ongoing crisis in Flint, Michigan.

Well-integrated BGI continues to gain policy traction as a safe, cost-effective, and more sustainable approach to managing flooding and stormwater pollution that provides smaller legacy cities with multiple eco-nomic benefits and ecosystem services. BGI typically

The Licking River Wastewater Treatment Plant in Newark, Ohio, saves nearly $14,000 per month in electricity from the installation of new diffusers and blowers and from overall increased capacity, allowing the plant to accept new waste streams for added revenue. *Source: City of Newark Wastewater Staff.*

includes multiple landscape treatments (such as pervious pavements, street trees, and pocket parks) that leverage natural ecosystems to reduce destructive stormwater impacts. These projects can act like sponges, soaking up rain and reducing demands on gray stormwater infrastructure. This in turn reduces the likelihood of wastewater overflow, which can flush pollutants (heavy metals, sediments, and nutrients) into local and regional bodies of water, diminishing water quality. BGI's value has become so widely recognized that the EPA recently authorized the inclusion of green infrastructure projects in federally required stormwater management plans (US EPA 2022b).

BGI, in tandem with gray infrastructure upgrades, can further provide both environmental and economic benefits for smaller legacy cities while safeguarding upstream watersheds. These changes might include stewarding and capitalizing on rich water assets, attracting private investment, providing construction and maintenance employment, and offering workforce training. Smaller legacy cities can, for instance, learn a great deal from Newark, Ohio, a nonlegacy city of 50,000 that converted a downtown sewer upgrade project into a revitalization opportunity that generated economic growth by leveraging multiple public funding sources (Elam 2019).

LOW-IMPACT DEVELOPMENT PROGRAMMING

Low-impact development (LID) programs for managing stormwater minimize the built environment's effects on water supply and cleanliness and have become a common local strategy for implementing BGI. The EPA describes this approach succinctly: it "treats stormwater as a resource rather than a waste product" (US EPA 2022c).

Local governments often establish an LID program within a public works, building, or planning department, providing education and technical assistance for homeowners, businesses, and developers. Programs might include homeowner workshops and training on how to manage stormwater, or grants and other financial incentives for homeowners and businesses to adopt LID practices. More cities are also adopting regulations that impose impact fees on businesses and homes based on their properties' percentage of impervious surfaces, which exacerbate stormwater runoff. Smaller legacy cities should also consider adopting and funding comprehensive LID ordinances that prioritize climate resilience and elevate equity and community stewardship through design and maintenance.

Enacting a comprehensive LID ordinance often starts with an inventory of existing local development and engineering codes to identify barriers and ensure that new codes apply to an appropriately wide range of current and proposed uses (US EPA 2021d). An LID ordinance should also guide greening existing and future public and private infrastructure like roads, parking lots, medians, curbs; and street trees; additionally, it should incorporate ideas like Smart Growth America's "Complete Streets Guidelines" for integrating LID and green infrastructure (National Complete Streets Coalition n.d.). Other important aspects of LID ordinances include enforcement fees to discourage features that promote stormwater runoff and financial incentives to encourage green infrastructure practices and landscape treatments.

Legacy cities of all sizes find these programs doable. Philadelphia's "Green City, Clean Waters" approach represents one of the most comprehensive local government efforts to integrate green infrastructure and LID throughout an entire city (Stutz 2018). The program represents a 25-year commitment, led for 15 years by the city's water department, with several principles and provisions embedded in the city's comprehensive plan and strengthened by ordinances, financial incentives, and technical assistance programs. Nonprofits,

the city's Office of Sustainability, and the public schools act as strategic partners in expanding the reach of LID practices and demonstration projects.

In Upstate New York, Monroe County and its seat, the City of Rochester, created a manual focused on LID retrofitting that many surrounding cities and towns have since implemented (Barton and Loguidice, DPC 2016). Regional approaches like this make perfect sense in smaller legacy cities, which often do not have the resources and technical capacities to develop their own LID or BGI plans from scratch. Access to nearby expertise and support can make all the difference.

VACANT PROPERTY RECLAMATION

Many smaller legacy cities already have robust urban greening initiatives that focus on reclaiming vacant properties for community gardens, urban agriculture, side yards, and pocket parks (Heckert et al. 2016; Mallach 2018b). In addition to complementing BGI and LID investments, this programming can generate a wide range of benefits and serve as a first line of defense for cities adapting to climate change and cultivating community resilience. If the land is put to thoughtful community use, the benefits include

Youngstown, Ohio, has taken a multidimensional approach to handling vacant and abandoned properties through the Youngstown Neighborhood Development Corporation, which has organized volunteer cleanups, created a Neighborhood Action Plan (co-led by the city) that engaged residents, and redeveloped vacant housing in partnership with the Mahoning County Land Bank. *Source: Youngstown Neighborhood Development Corporation.*

- **Economic**: increasing the value of adjacent properties, catalyzing neighborhood revitalization, and providing green jobs to build and maintain infrastructure;

- **Social**: enhancing social capital among residents, reducing neighborhood crime, rebuilding civic infrastructure, and providing opportunities for youth engagement;

- **Mental and Physical**: contributing to parks for recreation and active living, food security, and the calming effects of tree shade and other natural amenities; and

- **Ecosystem Benefits**: mitigating stormwater runoff, helping reduce urban heat islands, improving air quality, replanting with native plants and trees, providing animal habitat.

Legacy cities have reclaimed, demolished, and greened vacant properties for decades, and they continue to demolish thousands of abandoned homes, many of them victims of the 2007 housing foreclosure crisis. Subsequently supported by robust state and federal neighborhood stabilization initiatives, such efforts not only funded demolition by municipal and county governments but also generated sharper legal tools, technical assistance, and site reuse options that expanded neighborhood water filtration capacity (Shuster et al. 2014). As federal and state funding for intensive demolition work decreases, community-based organizations can continue related stabilization projects such as rehabilitation, repair, and support for low-income home ownership.

For smaller legacy cities, greening vacant lots can transform eyesores and liabilities into community assets, particularly in neighborhoods that historically have lacked access to green space, while also providing opportunities for community-led stewardship (Mallach 2018b). Green community development corporations, such as members of the Groundwork USA network, have a long track record of leading community-driven urban greening initiatives in smaller legacy cities, like brownfields remediation, pocket farms, urban gardens, riverfront restorations, parks and trails, and replanting native plant species (Groundwork Ohio River Valley n.d.).

Another example is the work of the Youngstown Neighborhood Development Corporation (YNDC). In just 10 years (2010–2020), YNDC transformed more than 300 vacant lots, rehabbed 130 vacant housing units, repaired 349 owner-occupied homes, and sealed up 2,342 vacant houses—all while providing an average of 64 jobs a year (Youngstown Neighborhood Development Corporation 2020). With its emphasis on all three of the greening policy areas, YNDC also provides the foundation for comprehensive green planning.

Legacy cities have reclaimed, demolished, and greened vacant properties for decades, and they continue to demolish thousands of abandoned homes, many of them victims of the 2007 housing foreclosure crisis.

County land bank authorities are often a critical partner and player in the greening of vacant properties throughout the Rust Belt. Authorized by state enabling law, the contemporary expansion of land-banking authorities originated in Genesee County, Michigan. Similar laws are now on the books in 16 other states, including Ohio (Brown 2015; Center for Community Progress n.d.). Over the past 20 years, county land bank authorities in areas that include legacy cities (such as Cuyahoga County, Ohio, and Genesee County, Michigan) have also become instrumental in demolishing vacant and often tax-foreclosed properties and then working with community development organizations to maintain and green vacant lots for pocket parks, green infrastructure, and community gardens (Brown 2015).

Smaller legacy city leaders can also consult several useful vacant lot greening guides published over the past decade (Lichten et al. 2017). Building on 2009's pioneering book *Re-Imagining a More Sustainable Cleveland,* many legacy cities have developed similar community resources, such as Baltimore's *Growing Green Initiative* (2014) and South Bend's *Lot Renewal Resource Guide* (2018). These each provide community-based organizations with details on site specifications, costs, tools, and techniques to convert vacant lots for stormwater management, energy generation, food production, tree farming, and more (Cleveland Urban Design Collaborative 2009).

Greening of Detroit has led community-based planting in Michigan's largest legacy city to restore the tree canopy and expand urban forests, which can help keep dangerously high "heat island" temperatures in check. *Source: National Association of State Foresters.*

URBAN FORESTRY INITIATIVES

Thanks to warming temperatures and poor maintenance of urban tree canopies, hundreds of cities across all regions of the United States now experience the urban heat island effect (figure 3.1), which happens when a city's pavement and buildings absorb heat from the sun and radiate it outward. This can dramatically increase the ambient temperatures in neighborhoods with fewer trees and water bodies. Heat has been the country's leading weather-related cause of death for three decades, especially for populations that lack green spaces or air conditioning (Khatana 2022).

Extensive research shows that communities of color suffer disproportionately from urban heat island effects. One study found that historically redlined sections of Baltimore, Maryland; Dallas, Texas; and Kansas City, Missouri, now have higher mean land surface temperatures than other areas of each city (Wilson 2020). In essence, the segregationist practice of redlining lives on in the current deficit of accessible green amenities and trees in communities of color.

In addition to combatting climate change, urban forestry programming protects existing trees and expands tree canopies to cool urban heat islands, sequester carbon, save energy costs, act as wind guards, help manage stormwater runoff, and provide natural habitat. Planting ecosystem-appropriate trees can reduce outside surface temperatures as much as 20 to 40 percent (US EPA 2008; Washington Nature 2019).

While municipalities have long handled tree management, they have increasingly adopted urban forestry master plans, established urban forestry departments, and hired urban foresters dedicated to preserving and expanding community tree canopies. For many smaller legacy cities, the parks and recreation or public works departments may manage trees, and existing work may focus almost exclusively on trees within public rights of way and on public property. The new generation of urban forestry practices, however, expands those responsibilities with a systematic approach that

Figure 3.1

Urban Heat Island Profile

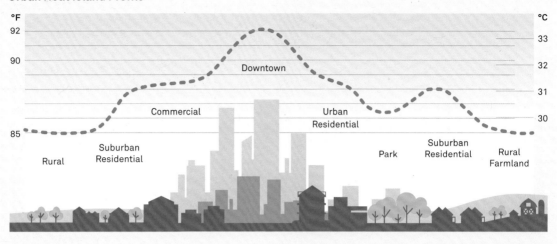

Source: Heat Island Group, Lawrence Berkeley National Laboratory.

involves extensive inventories and ecosystem analyses, including on private property with owner permission, with proactive policies and master plans (Davey Resource Group, Inc. 2018; Vibrant Cities Lab n.d.a). In addition to local governments' general funds, other urban forestry funding sources include permit fees, special assessment districts, and infrastructure and maintenance funds, along with a blend of government and philanthropic grants (Vibrant Cities Lab n.d.).

Community members often conduct inventories with support from national and regional GIS tree canopy mapping tools. Many local governments also partner with nearby universities and colleges, along with nonprofit environmental organizations, for technical assistance and expertise in urban forestry. Urban greening nonprofits may also have programs that engage neighbors and youth in removing invasive trees and creating micro urban forests with native species. Federal agencies such as the US Department of Agriculture Cooperative Extension Service and the US Forest Service provide relevant technical assistance and guidance to municipalities, in consultation with national environmental and natural resources groups (US Department of Agriculture 2016). New tools, such as the Forest Service Climate Change Resource Center's "i-Tree," even enable local governments to calculate

the economic and climate benefits that specific trees provide (US Department of Agriculture n.d.a.). Through the IRA, nonprofits, state and local governments, and community-based organizations can now also apply to the USDA for project grants totaling more than $1 billion to expand tree cover, ensure equitable access, and bolster climate resilience in Justice40-eligible communities (US Department of Agriculture 2023).

The midsize legacy city of Worcester, Massachusetts, is something of a smaller legacy city pioneer in urban forestry resilience. In 2008, the city was forced to cut down 25,000 trees to address serious beetle infestations, mainly in one of its poorest neighborhoods. With support from the Massachusetts Department of Conservation and Recreation, it then planted 30,000 new trees by 2014, many of them in stormwater catchment areas (US Department of Agriculture n.d.). Worcester also committed to hand-planting by local residents, rather than machine-assisted planting by contractors, and providing training and employment opportunities in the afflicted communities.

Following the success of this project, the state launched similar tree programming for its smaller legacy cities, called "Greening the Gateway Cities" (GGCP). Supported by four state agencies and local

The Capital Trail hiking area in Richmond, Virginia, received recognition from then-Governor Terry McAuliffe in 2016 following the Capital Trees partnership to improve its green spaces through collaboration among the city, corporations, nonprofits, and individuals. *Source: Lisa Trapp/Capital Trees.*

community partners, GGCP targets neighborhoods with significant tree deficits and large rental populations for tree restoration and program employment. It also tracks neighborhood temperatures and energy use over time to determine the program's economic and environmental effectiveness (Commonwealth of Massachusetts n.d.).

Such data is also useful for more than urban forestry planning. For example, the City of Richmond, Virginia, sent out citizen scientists through its sustainability office and a few partners to collect more than 60,000 temperature readings to develop heat vulnerability maps. Data and analysis from these maps helped inform several planning processes, including a comprehensive plan update, implementation of the city's integrated watershed management plan (RVAH2O), and its equity-centered CAP (Hoffman and Dunn 2021).

Brownfields Redevelopment and Green Buildings

The manufacturing history of legacy cities presents opportunities for recycling contaminated land and for the adaptive reuse of former commercial and industrial building sites (called "brownfields") along with the potential for green jobs, ranging from environmental cleanup and building deconstruction to eventual retail, hospitality, or office work. Beyond the inherent sustainability of adaptive reuse compared to new construction, many of these properties can also be repurposed for light manufacturing and other low-carbon industries (Leigh and Hoelzel 2012).

Many smaller legacy cities already have extensive experience redeveloping vacant industrial and commercial properties that were once a critical part of their built environments. Over the past quarter-century, hundreds of local governments and their partners have relied on public brownfields programs led by the EPA's Land Revitalization Program to remove or contain the toxic hazards found on industrial, commercial, and other lightly-to-moderately polluted properties. The 2021 Infrastructure Investment and Jobs Act dramatically increased funding for the EPA's signature Brownfields Program to $1.5 billion over five years, an annual increase of $100 million, with $1.2 billion for competitive grants and $300 million for state and tribal categorical grants (US EPA 2023).

Dozens of legacy cities, such as Cleveland, Ohio; Lowell, Massachusetts; and Pittsburgh, Pennsylvania, were also among the first and second waves of municipalities to establish local brownfields redevelopment policies and programs that have catalyzed revitalization in their downtowns, riverfronts, corridors, and neighborhoods. Overall, both federal and state brownfields projects have generated significant economic and environmental benefits to the projects themselves, adjacent properties, and communities at large (US EPA 2019).

Given this recent history, the scaling of existing brownfields redevelopment programming offers smaller legacy cities a great opportunity to look at ways climate resilience, environmental justice, and green economic development intersect with existing infrastructure.

- **Climate Change**. In response to climate change, the EPA's land revitalization agenda has broadened the range of reuse options it encourages local government to pursue, including green infrastructure, renewable energy generation, and long-term recovery from natural disasters. After years of careful research and pilot testing, these programs are described in the agency's *Climate-Smart Brownfields Manual* (US EPA 2021a).

- **Area-Wide Planning**. Unlike individual property-level cleanup and renovation, area-wide brownfields redevelopment encompasses the coordinated, cohesive cleanup and reuse of multiple sites. In 2010, the EPA launched its competitive Brownfields Area-Wide Planning Pilot Grants, which enabled local governments to test how communities could remediate, reuse, and prioritize multiple sites within a specific district, corridor, or neighborhood. Groundwork USA subsequently released related recommendations for local government and community leaders, emphasizing smaller legacy cities such as Lawrence and Chicopee, Massachusetts, and Janesville, Wisconsin (Fowler and O'Brien 2017; US EPA 2010a; US

The open hearth stacks from US Steel's Homestead Works in Pittsburgh, Pennsylvania, still stand to pay homage to the complex's former use as a steel mill. *Source: Sean Pavone/iStock/ Getty Images Plus.*

Area-wide brownfields planning in Lawrence, Massachusetts, advances the community's vision with cleanup and reuse projects along the banks of the Spicket River. *Source: Groundwork USA.*

EPA 2021b). Based on lessons from these 23 pilot grants, EPA developed its current multi-purpose brownfields grant program. At the state level, New York's Brownfield Opportunity Area program offers an even stronger model for how to create area-wide brownfields redevelopment plans, as it provides municipalities with funding that covers environmental assessments, cleanups, and planning along with technical assistance (New York State n.d.).

- **Brightfields**. Brownfields also provide opportunities for reuse plans that feature renewable energy or "brightfields." The RE-Powering America's Land Initiative, a 2008 EPA partnership with the US Department of Energy, has mapped over 100,000 potential sites, including capped landfills and mines, on more than 44 million acres, and assisted in 417 installations producing 1.8 gigawatts (GW) of electricity (US EPA 2021c). The 2022 IRA (IRA) also includes new tax credits, additional funding, and technical assistance opportunities for localities (Popkin 2022; White House 2023).

- **Brownfields Job Training**. Part of EPA's brownfields portfolio also includes grants for nonprofits, colleges, and local governments to recruit, train, and place unemployed and underemployed residents in entry-level jobs doing brownfields cleanup, containment, and monitoring. Program

graduates gain the skills (and often certifications) needed to secure full-time, sustained employment in various aspects of hazardous and solid waste management; they also learn about reuse strategies and how to further their careers in related employment sectors (US EPA 2022d).

- **Environmental Justice (EJ)**. The EPA's Office of Environmental Justice (OEJ) collaborates with national and local environmental justice organizations to expand residents' access and power in selecting, designing, and developing brownfields so they can benefit from their reuse. As part of its broader Thriving Communities Initiative, the EPA and the Department of Energy launched a national network of 17 regional technical assistance centers in 2023 to provide underserved communities with EJ training and capacity building, convene and engage with local community groups, and expand access to federal resources (US EPA 2023a). Through grants from the US EPA, Groundwork USA offers additional equitable development technical assistance and peer-support programs through its own national network, customizing inclusive planning activities and redevelopment agendas for brownfield-affected communities (Fowler and O'Brien 2017).

Building Out Renewable Energy Infrastructure and Equity

With plentiful brownfields and grayfields (underused or vacant commercial properties, with little or no contamination) and proximity to significant water bodies and rural land, smaller legacy cities can derive economic benefits from generating, distributing, and manufacturing renewable energy infrastructure. Coupled with federal, state, and private investments, legacy city leaders are well positioned to build upon their industrial heritage by modernizing their energy infrastructure and creating careers in its management, manufacturing, construction, installation, and maintenance.

Beyond generating lower GHG emissions and jobs, renewable energy enables smaller legacy cities to rectify issues of environmental injustice and inequity; many of their Black and Brown communities have historically suffered (and continue to suffer) from the deliberate, excessive number of power stations and toxin-generating industries located in or near their neighborhoods. Remedying these trespasses must be a top priority in revamping existing energy infrastructure for a more equitable, climate-resilient future, and smaller legacy cities can and should proactively design, deploy, and distribute efficient renewable energy resources accordingly.

RENEWABLE ENERGY GENERATION AND DISTRIBUTION

Most US electricity comes from investor-owned regional "utility-scale" facilities, which require long transmission lines to reach local distribution networks that then connect to end users. Renewable energy can be generated at utility scale, as with solar and wind farms, but it can also function on a much smaller scale, feeding directly into an individual structure (as with rooftop solar panels) or into local distribution networks dubbed the "microgrid" (Cleary and Palmer 2022; US EPA 2015). How electricity moves from a

---

Groundwork Lawrence helped transform a brownfield site in Lawrence, Massachusetts, into the Spicket River Greenway, a 3.5-mile trail offering renewed views of and access to the river. *Source: Adi Nochur/Groundwork USA.*

Community-based
organizations in Toledo,
Ohio, partnered to develop
a 21,000-panel solar field
on the site of a former
Jeep manufacturing plant.
Source: Toledo Aerial Media.

generation source to its end use determines what local officials can and cannot do to build out renewable infrastructure.

A small but growing number of municipalities own and operate energy facilities as public utilities or belong to a community-choice aggregation plan; otherwise, the policy levers available to local officials to develop the benefits of renewable energy, particularly in small cities, differ quite dramatically, depending on distribution grids (Kennedy 2021; Perkins 2019). Indeed, utility-scale renewables commonly involve grid challenges that lie beyond the control of most municipal leaders and are more federal, state, corporate, and technological in scope, though not entirely.

Although most legacy cities have large swaths of vacant land, it may not be enough to support utility-scale energy generation. Also, large solar or wind turbine installations are usually incompatible with the compact urban form of many smaller legacy city neighborhoods, which itself supports low-carbon urban design. Urban energy sitings can also raise neighborhood conflicts and environmental justice concerns.

For legacy cities, then, strategies that develop distributed renewable energy, such as via community solar programs, are much more appropriate. In this model, consumer subscribers purchase "shares"

in off-site solar farms owned by private utilities, third-party developers, or community-controlled entities. They are then billed or credited monthly based on share amount. These arrangements result in lower monthly bills, require no upfront costs, are available to renters, and do not require homeowners to install solar panels—making them an accessible, advantageous option in many communities.

Though these arrangements can operate at utility scale, ideally, community solar in urban areas leverages brownfields, surplus government land, or abandoned shopping centers, warehouses, parking lots, and storage facilities. Public rights of way—namely, land adjacent to local roads and even state highways—offer another siting opportunity for renewable energy generation. For example, in Toledo, Ohio, community-based organizations partnered to develop a 21,000-panel solar field on the site of a former Jeep manufacturing plant (e.g., brownfields) with support from the Greater Toledo Community Foundation. The site now sells energy to nearby companies and uses the profits to finance low-carbon energy projects in nearby low-income neighborhoods (Greater Toledo Community Foundation 2021).

State regulations and public utility laws can either facilitate or hamper these projects, however. At the time of publication, 22 states plus Washington, DC, have community solar–enabling laws, but most projects are new or still in the pipeline, so their progress is difficult to assess. Until this legislative gap is filled elsewhere, pro-solar states such as Ohio will miss out on viable renewable energy uses for vacant properties, and residents will forfeit potential significant savings and climate mitigation opportunities (Welter 2021).

The right mix of local political savvy and regulatory streamlining can make distributed renewable energy even more hyperlocal. With a small 2005 state grant, volunteers in Ypsilanti, Michigan (population 21,000), formed SolarYpsi, a grassroots initiative that installed a four-panel solar array at the local food co-op. In response, the city took additional steps to facilitate solar, reducing the soft costs of installation by streamlining permitting and inspection, removing development roadblocks, and staying abreast of low-carbon tech market trends.

The municipality also worked with SolarYpsi to cultivate enthusiastic public support. As a result, by 2017, Ypsilanti had received a Gold designation from SolSmart, a Department of Energy–funded program recognizing local solar power development. That demonstrates how organizations such as SolarYpsi can be both instigators and partners in smaller legacy cities' renewable energy and other climate resilience planning (Energy News Network 2018; Environment America 2019; SolarYpsi n.d.).

Utility-scale power projects can also link local economic development initiatives to state planning, which large projects require anyway. The wind energy infrastructure in New Bedford, Massachusetts, offers a case in point: The former whaling and textile industry hub remains the country's largest commercial fishing port in dollar value of catch, and its downtown New Bedford Whaling National Historical Park, anchored by the New Bedford Whaling Museum, has helped expand the local arts, tourism, and hospitality industries (National Park Service 2022). Still, it has remained a struggling midsize legacy city.

In 2009, the state established the Massachusetts Clean Energy Center (MassCEC), a quasi-governmental economic development agency that promotes renewable energy, green jobs and buildings, and clean transportation. MassCEC also commissioned and manages the New Bedford Marine Commerce Terminal, a renovated brownfield engineered to deploy and marshal wind turbines and other components for three major commercial offshore wind farms in federal waters south of Martha's Vineyard and Nantucket, the first of which is scheduled for completion by the end of 2023 (Asimow and Kushwaha 2022).

New Bedford stands to benefit from the project, as will those who receive specialized workforce training and union jobs. The estimated statewide economic impact includes 9,850 jobs over 10 years and $2.1 billion added to the Massachusetts economy. To comply with state GHG mandates, these wind farms will generate enough energy to power a third of the state's residences by 2027. These projects also follow

SolarYpsi volunteers install solar panels atop a fire station in Ypsilanti, Michigan. *Source: ESAL.*

other state laws that require utility companies to buy offshore wind power and that increase offshore wind generation goals, expand on energy-efficiency standards, and strengthen environmental protections.

New Bedford's Economic Development Council, with a staff of just seven people, has had a strong hand in these and other successes. In the early stages of wind infrastructure planning, it housed the New Bedford Wind Energy Center (WEC), a dedicated internal division that coordinated governmental, private sector, and nonprofit resources. As the work matured, the WEC was subsumed by the New Bedford Port Authority and then by the New Bedford Ocean Cluster, established in 2017 to support wind, fishing/processing, aquaculture, and other maritime activity focused on innovation and technology. In 2021, the city's Redevelopment Authority executed development deals on two additional waterfront building projects: a retrofitted workforce training center now owned and managed by Bristol Community College (with EPA funding to support hiring and training in

underserved local communities) and a staging and logistics facility for regional utility companies on the site of a long-decommissioned power station (Froias 2022; Lennon 2021). City officials and the New Bedford Economic Development Council have played a critical formal role in the entire arc of New Bedford's emerging wind energy industry.

With strong assistance from the state over two gubernatorial administrations, the offshore wind project is finally coming together, demonstrating the importance to smaller legacy cities of state-level continuity in green economic development policy.

US Senator Edward Markey speaks at the July 2021 signing of a project labor agreement between Vineyard Wind and the Southeastern Massachusetts Building Trades Council in New Bedford, Massachusetts. *Source:* The New Bedford Light.

ENERGY EFFICIENCY

A chief feature of green buildings, energy efficiency remains one of the more common and practical elements of local GED strategies. When properly incentivized by state and local regulations and executed by energy nonprofits and the private sector, it has two enormous benefits: Energy-efficient homes and businesses reduce GHG emissions and energy costs and generate jobs in manufacturing, installation, maintenance, and retrofitting—which smaller legacy cities with older building stock frequently need. Renewable energy subsidies also usually require energy audits, which can help resolve leaks and other issues. Newer resources to expand residential energy efficiency programs through the IRA will also advance climate equity and resilience among homeowners and renters of color (Ungar and Nadel 2023).

Most legacy cities already have experience with adaptive reuse of older industrial and commercial buildings that could help support a local energy efficiency industry, and they also tend to maintain strong building, construction, and repair workforces. With the right investments in energy efficiency, properties in smaller legacy cities could provide job training for local workers to learn how to construct, retrofit, and maintain eco-friendly buildings. With emerging federal and state energy resources and assistance, too, smaller legacy cities can collaborate with regional economic development agencies and alliances to invest in retooling small manufacturing firms for the broader energy efficiency supply chain.

If state law permits, smaller legacy cities should adopt a green building code to establish baseline energy efficiency standards. Some states require local governments to follow their standards for green buildings while others enable local governments to adopt their own. Several professional associations—of which Leadership in Energy and Environmental Design (LEED) is the best known—offer green building models, although these mainly focus on new construction or adaptive reuse of brownfields, not on rehabilitation

of aging homes and buildings. LEED standards for buildings remain the most common model codes for most US local governments. LEED codes also cover standards for new green building design and materials, climate resilience, water consumption, LID and green infrastructure, transportation, energy-efficient consumer products, and alternative energy generation.

With the right investments in energy efficiency, properties in smaller legacy cities could provide job training for local workers to learn how to construct, retrofit, and maintain eco-friendly buildings.

For retrofits, however, LEED building standards apply only to operations and maintenance. Owners retrofitting older houses must rely on specialized, licensed contractors, although some state and federal government programs subsidize this work. Given that many landlords do not pay their tenants' energy costs, local officials working in code enforcement and licensing should become familiar with applicable incentive programs and strategies that might help renters in the future.

A new, fast-evolving program offers indispensable counsel during this period of low-carbon transition: the Energy Equity for Renters initiative. Designed by the Urban Sustainability Directors Network and the American Council for an Energy-Efficient Economy, the initiative offers a research and action guide, a toolkit for local government officials, and a regularly updated policy tracker (ACEEE 2023). Beyond building codes, smaller legacy cities should further invest in educational resources and technical assistance to help developers, businesses, and especially homeowners put these codes to use. Local officials can enlist relevant associations of home builders and developers, as well as seek partnerships with local community colleges and workforce development boards, all of whom can recruit green contractors to install and maintain green building systems.

ENERGY EQUITY

Legacy cities' low-income residents and communities of color typically have greater "energy burdens" than their wealthy or white neighbors, spending a disproportionally higher share of their more limited monthly incomes on energy costs. In fact, low-income households spend an average of 8.1 percent of total income on energy costs, compared to 3.1 percent for median household use (Pelton and Kane 2021). Reducing this disparity is both fair and climate smart, as home energy use contributes nearly 20 percent of the country's heat-trapping greenhouse gas emissions (Goldstein et al. 2020; Pelton and Kane 2021; US Department of Energy 2018). The COVID-19 pandemic further exacerbated these energy burdens because many low-income residents and people of color lost their jobs or had to stay home, per state and local lockdowns.

In cultivating actions for both public and private energy efficiency, local governments have a range of resources to develop that covers climate resilience, equity, and GED programming. Elevate Energy, a nonprofit in Chicago, for example, engages Great Lakes–area consumers and communities interested in energy efficiency and controlling energy costs, with an emphasis on equity. Its portfolio includes work with high-performance buildings, solar energy, and energy efficiency, and it bridges the interests of private energy firms, local governments, and community-based organizations. Elevate Energy also conducts applied research on statewide energy-generation policies, along with case studies of promising practices in Chicago and beyond (Elevate Energy 2023). Additionally, several projects focus on the neighborhood level and on job training for individuals and contractors. They also provide classes on the market value of solar energy and energy efficiency in residential buildings for real estate agents and appraisers (Brookstein and Savona 2019).

Energy equity programing deliberately ensures low-income households are "weatherized" for efficient energy consumption, have access to well-paying jobs in the low-carbon energy sector, and enjoy the benefits of renewable energy—including the expansion of its use. Local governments are largely beholden to state policy and the private sector in meeting these objectives, but they can effectively channel resources and effort toward meeting equitable goals, especially with the help of nonprofits like Elevate Energy that are deeply versed in the financial, technical, and policy landscape.

Elevate Energy, based in Chicago, Illinois, partners with communities and consumers throughout the Great Lakes region to improve energy efficiency, foster energy equity, and provide job training opportunities. *Source: Elevate.*

CHAPTER 4

Getting Started with Greening

As most local officials know, good planning and administration require initial assessments of conditions and capacities on the ground, against which progress over time—or the absence thereof—can be compared. Understanding baseline conditions, inventorying assets, and establishing core capabilities is also crucial for coordination among municipal departments, regional partners, and funders and to ongoing political accountability and healthy community engagement. With such foundations in place, local governments can effectively expand existing sustainability initiatives and take full advantage of federal, state, and philanthropic sustainability resources.

EPA Administrator Michael Regan (left center) and Jackson, Mississippi, Mayor Chokwe Antra Lumumba (right) listen as fourth-grader Kingston Lewis talks about a lack of drinkable water at Wilkins Elementary School in November 2021, a problem that forced several schools in Jackson to resort to temporary virtual learning. *Source: Rogelio V. Solis/AP Photo.*

Offices of Sustainability

Several smaller legacy cities already have part-time sustainability coordinators or managers charged with aligning multiple departmental missions with common green practices. Often housed in environmental services, public works, or planning departments, local sustainability staff have, however, faced increased demands on their time as the urgency of climate change, racial injustice, and wealth and income inequality have expanded in many states. With the new infusion of infrastructure resources in the IIJA and the climate and energy programming in the IRA, together providing the United States' first significant climate funding package, city sustainability staff have more to juggle than ever.

To gain access to federal, state, philanthropic, and private sector funding for transformative green regeneration, local governments need to expand their capacities for grant writing, managing multiple programs, growing networks, and enhancing programming expertise. Smaller legacy cities are usually caught in a catch-22, however: they lack the capacity to expand capacity. Local officials can take a number

Since 2019, Lynn, Massachusetts, has been home to "The Resident," a mural by Smug One created through the Beyond Walls program to introduce public art to smaller legacy cities. *Source: Philip Fagan/ Beyond Walls.org.*

of approaches to unfreezing this capacity standoff, but establishing a sustainability office with a full-time director is key to setting up third-generation sustainability initiatives (Reale 2021).

The infusion of federal climate and infrastructure resources can help cities to hire a full-time sustainability coordinator, with explicit responsibility for scaling current sustainability initiatives, identifying funding streams, and coordinating with other departments and with state or regional programs. With sufficient political support, the coordinator can eventually set the stage for a full-time director and dedicated office or department that can better align city agencies and marshal critical partnerships that integrate climate resilience and equity with GED. A dedicated office should provide core capacities such as

- coordinating policies and programs across departments that can align programming on climate resilience, equity, and GED;

- crafting a compelling green regeneration message around pressing priorities, such as generating more and better-paying jobs, upgrading deteriorating water and transportation infrastructure, creating healthier, detoxified neighborhoods, lowering energy and healthcare costs, and reducing urban heat islands;

- engaging with the community more equitably and responsively to elevate and empower local residents as sustainability stewards, especially in transitioning away from fossil fuels and in neighborhoods where environmental burdens on people of color are most acute;

- communicating current sustainability activities to diverse audiences, anchored by well-designed, accessible web platforms, with effective social media capacity and relevant data dashboards that measure progress among key indicators;

- collaborating and strengthening relationships with local community partners—such as regional entities, local universities, nonprofits, businesses, financial institutions, and community-based organizations—and with regional, state, and federal agencies (discussed in chapter 5); and

- conducting baseline policy screens to assess climate risks, racial disparities, and existing sustainability assets with support and participation from community volunteers, when appropriate.

Several midsize legacy cities and counties, such as Richmond, Virginia; Providence, Rhode Island; and Cuyahoga County, Ohio, have fairly well-established sustainability offices, each with a full-time director and one or two staff who can help manage new and existing sustainability programs and grant opportunities along with external activities (Schilling and Velasco 2020). The Richmond and Providence sustainability offices were instrumental in developing and implementing their respective citywide sustainability and climate equity plans.

Midcareer Sustainability and Regeneration Fellowships

Multiyear fellowships for experienced professionals can expand municipal capacity, fill knowledge and skills gaps among current staff, and prepare fellows for permanent roles while minimizing personnel turnover (Schilling, Hexter, and Bulka 2015). The Detroit Revitalization Fellows program for midcareer professionals, housed at Wayne State University with major foundation support, is an excellent model that leads to permanent job placement after two years. (The program was on hiatus at the time of publication, while its staff went through a strategic planning process to align programming with the city's changing needs.) (Wayne State University 2023). Another model is the Massachusetts Transformative Development Initiative (TDI) fellowship program, administered by the state's economic development agency. TDI's three-year midcareer fellows provide technical assistance and strategic community development guidance to targeted districts in the state's smaller legacy cities (MassDevelopment n.d.). Both programs pay competitive salaries and benefits.

As part of its funding from the 2020 Coronavirus Aid, Relief, and Economic Security (CARES) Act, the US Economic Development Administration launched a similar program in 2023: the Economic Recovery Corps (ERC). Through it, the International Economic Development Council, in collaboration with coalition members including the National League of Cities and International City/County Management Association, will place, manage, and train 65 midcareer fellows to work in disadvantaged communities for three-year terms with host organizations, pursuing transformative green economic development policies, programs, and projects (US Economic Development Administration 2022).

Climate and Equity Screens and Inventories of Overlooked Physical Assets

Long-term greening strategies require in-depth examination of existing programs and sustainability initiatives in both customary and emerging urban policy areas, including housing, economic development, land use, the environment, workforce, transportation, energy, and climate resilience. Such a review necessitates inventorying and screening the community's built environment to determine conditions on the ground and find programming gaps. This information can then be converted into indicators and benchmarks to circulate among municipal staff, community members, funders, and other key stakeholders (Feiden and Hamin 2011).

Inventories of physical liabilities and assets are labor intensive but useful for measuring hard stock such as vacant properties, tree canopy, regional-level greenhouse-gas-emitting activity, and other data sets

The State of Rhode Island and its capital, the City of Providence, have both committed to reaching major clean energy milestones by 2030. *Source: Climate Action Rhode Island.*

(Bloomberg Cities 2020). They also establish a baseline for developing action plans with strategic goals, indicators, and methods for future program evaluation.

Screening, by contrast, involves reviewing plans for specific content that is often missing from such documents. Screening tools usually take the form of checklists or matrixes (which weight as well as list each entry) and sometimes draw from survey data. Though relatively new, screening for climate and environmental justice is increasingly common in urban planning, with a growing number of tools available (Geertsma 2019; Resilient NJ n.d.).

Climate screens pinpoint and prioritize various climate-related disaster risks, from flooding and fire hazards to urban heat exposure. As such, screens are important for developing meaningful vulnerability assessments and adaptation plans and are increasingly designed for concurrent use, as with the Biden Administration's Climate and Economic Justice Screening Tool (White House 2022). Once generated, these data should be regularly updated and shared across departmental silos, with other cities in the metropolitan area, within a city's various communities, and, ideally, through an online municipal data dashboard. Such findings have broad implications for communities of color, people experiencing poverty, seniors, children, and anyone confronting the daily impacts of climate change.

Equity screens further identify acute sources of environmental inequity and may be a good first step in changing how responsive policy develops. Although they often lack enforcement, robust standards, or accountability mechanisms, equity screens are themselves important educational tools, and using them can signal a cultural shift in local government.

The federal EPA offers a good equity screening tool, as do individual states such as California and Maryland (California Office of Environmental Health Hazard Assessment n.d.; Maryland Department of the Environment 2022; US EPA 2023b).

Local officials and activists can also lobby for state laws that support climate and environmental justice planning. New York State's 2019 Climate Leadership and Community Protection Act (CLCPA), one of the country's most aggressive state climate laws, sets a legally binding emissions reduction standard and mandates that 40 percent of state climate and energy funding be invested in environmental justice communities (New York State 2023). NY Renews led the development of the legislation with a statewide coalition of over 300 labor unions, community groups, faith communities, and environmental justice advocates, including from the legacy cities Albany, Binghamton, Buffalo, Rochester, Schenectady, Troy, and Utica (NY Renews 2022).

Exploratory Scenario Planning: Envisioning a Slippery Future

Exploratory Scenario Planning (XSP) can be a useful tool for community engagement, problem solving, and consensus building amid the many uncertainties of climate change. Devised by post–World War II military strategists and later tailored to business organizations, scenario planning began generally as a group process for identifying choke points or obstacles to achieving a "normative" goal. *Exploratory* scenario planning works in reverse: It first proposes three or four outcomes and then identifies developments that could disrupt each outcome, indicators that signal coming disruption, and mitigation measures that might minimize or derail disruptive change (Futrell 2019; Stapleton 2020).

As XSP explores plausible futures in which past trends do not necessarily carry predictive value, there is no single correct approach to the issues on the table—which is why in practice the process is primarily collaborative and educational rather than contentious. Conducted by a neutral facilitator over several workshops, the process begins with a focal question such as how to plan for climate in-migration (Rajkovich et al. 2022). A variety of "scenario narratives" are then discussed and evaluated by workshop participants, who may include municipal department heads, representatives of stakeholder groups, and subject-area experts. Although too large or broad a participant pool can stymie proceedings, public access to planning discussions and debates (such as through video, print material, and special public events) critically engages the community with the complexities local officials face in guiding residents through the uncharted territory ahead—and ensures underrepresented voices are heard along the way (Stapleton 2020).

A key part of CLCPA mandates that state agencies apply equity and climate screens to regulations for which they are responsible, which can entail activities like identifying ("mapping") targeted communities, developing and tracking metrics of progress, and requiring agencies to promulgate certain related regulations and to involve meaningful public input. This kind of formal enforcement is a first step that other states with high concentrations of legacy cities like Michigan, New Jersey, Ohio, and Pennsylvania could replicate (Environmental Advocates of New York 2019; Sustainable Jersey 2020).

Equitable Community Engagement

As seen in chapter 2, environmental justice activists have become wary of relying too heavily on governments, courts, and the limited aims, tools, and regulatory reforms of the past. They have learned to supplement those interventions with actions that seek to transform systems in a holistic manner, rather than through incremental or siloed approaches (Karner et al. 2020). Environmental justice (EJ) therefore demands that local governments adopt new approaches that redistribute power, facilitate community ownership and neighborhood stewardship, and invite new contributors into the city's process with great care and clarity of purpose (González and Facilitating Power 2019). Bringing in communities long plagued by environmental injustice is delicate and demanding, and effective engagement requires authentic commitment to building trust. Engaging existing grassroots community organizations is crucial to that effort, and community leaders must be granted access to the resources necessary to effect transformative solutions and drive policy. Transferring power to community members in this way is a vital start to building trust and public support for project implementation.

Here, local officials must provide resources, support, and useful knowledge while respecting independence and learning from community expertise—goods they might have trouble delivering and that are not always in harmony with traditional city processes and procedures. That said, they can take a number of proactive community-engagement actions, detailed in table 4.1, to expand access, increase transparency, and foster equity (Southeast Sustainability Directors Network n.d.).

Bringing in communities long plagued by environmental injustice is delicate and demanding, and effective engagement requires authentic commitment to building trust.

We also recommend learning from other cities. For example, Providence, Rhode Island, and Richmond, Virginia, have put together stellar climate equity plans by adopting and piloting many of the equitable community engagement practices discussed here. Other EJ communities, including those in Southwestern Pennsylvania and in Dilles Bottom, Ohio, organize virtual public "toxic tours" that could supplement in-person local visits (Toxic Tours 2022).

In Rhode Island, Providence's 2019 Climate Justice Plan offers a strong model for both community engagement and a final product (City of Providence 2019). This midsize legacy city and two nonprofit organizations created the city's first Racial and Environmental Justice Committee (REJC) in 2016, to integrate voices and concerns of people of color directly into the city's sustainability and resiliency planning. The 10-member committee includes representatives of various "seats" within the community, as well as five city representatives.

Table 4.1

Features of Equitable Community Engagement

Local government activities	Community participants	How does it work?	How does it benefit the community?
Respect community expertise and experience	Impacted community experts	Community experts coordinate division of labor in decision-making with local government officials by participating in review, inventory, and screening processes.	Pathways to policy and procedural influence are open, ensuring impacted communities have a direct say in the decisions that affect them, and that their on-the-ground expertise and lived experience are taken seriously.
Support grassroots organizing to strengthen community networks	EJ community organizers	Local government can channel funds to community organizations, support leadership training, and cede power in planning and programming.	Existing community efforts to provide governance, mutual aid, and other services can be supported and leveraged rather than replaced or duplicated.
Expand access to participation in public processes	Community members, especially task force or advisory board members	Local government offers engagement opportunities at flexible times, in convenient and accessible neighborhood locations, and with childcare, food, and other support such as stipends or similar compensation.	Planners and other local officials make it easier for constituents to meaningfully participate in public meetings, task forces, advisory boards, and other events and processes by prioritizing access and sustaining longer-term engagements.
Establish an environmental justice charter (Coming Clean, Inc. 2021)	High-level community leaders	With local government, environmental justice and grassroots communities, nonprofit and business leaders, and others clarify goals, organize priorities and values, and define a shared vision.	Though lacking regulatory or statutory power, a charter can foster shared analysis, unite like-minded community members, and lay the foundation for subsequent citywide policy and systems changes.
Launch participatory budgeting (Georgetown Climate Center n.d.)	Community members	Local officials involve the whole community in selecting priorities, responding to frontline needs, and investing in new interventions.	Local officials have a framework to respond effectively to on-the-ground issues and community members have a voice along the way, rather than being treated as an afterthought near the end of the process.
Visit places exposed to toxic chemicals and other impacted areas	EJ leaders, tour guides, and impacted community members	Planning and zoning board members, as well as other local officials, participate in professional site visits with guidance from impacted local leaders or official "Toxic Tour" guides.	Decision-makers witness urgent community needs firsthand, with direct input on remediation needs and healthcare interventions from those affected.
Apply "Targeted Universalism"	Diverse community members	All concerned demographic groups establish universal goals and targeted strategies to achieve them through cross-cultural collaboration that takes into account how each group is situated within political-economic structures and across geographies.	This approach, developed by legal scholar john a. powell, offers a unifying framework for those committed to equitable development from a variety of backgrounds and perspectives (powell et al. 2019).

The REJC was tasked with building bases of support within each community district, identifying key environmental justice issues and needs in the city's communities of color, and recommending a long-term process and structure for collaboration between communities of color and the city's Office of Sustainability. The committee proved instrumental in developing Providence's subsequent, widely admired 2019 Climate Justice Plan, as well as in the Office of Sustainability's adoption of its "Recommendations for a Just and Racially Equitable Providence" (City of Providence 2017). Providence's commitment to shifting power from institutional employees to impacted communities is unique among legacy cities—and across all US cities (PD&R Edge 2022).

Several critical elements led to the committee's success (City of Providence 2017). The REJC was BIPOC-centered and led, and it engaged consultants to conduct effective "Undoing Racism Trainings" across communities and among city government staff. Importantly, committee members employed by the city participated in the committee's efforts primarily by listening, and they conducted their own base-building at city hall to help other government officials engage in racial equity work.

In Virginia, the award-winning Richmond 300 plan offers smaller legacy cities another good model in content and format. It also elevated the voices of Latinx, Black, and low-income residents through extensive community engagement and provided a public process for shared analysis of how the community could confront the city's racist legacies (American Planning Association 2021; City of Richmond 2020).

The "First Floor Fund" of Dayton, Ohio, invests millions of dollars in retail and other small businesses that struggled to secure traditional financial loans and other funding as a result of the COVID-19 pandemic, with a focus on growing neighborhood commercial areas like the Historic Oregon District. *Source: Tom Gilliam.*

Small Business Development

Small business development presents an important baseline project for GED, and cities can help position entrepreneurs to launch and grow. Small businesses may not seem inherently "green," unless they provide green products or services or green their operations. However, they can be important conduits to closing the wealth disparities so prevalent in legacy cities—a vital aspect of greening—so access to capital, mentorship, and market support makes a difference.

Small businesses also provide jobs: those with fewer than 100 workers employ a third of the overall US workforce, a figure that includes far more prosperous metro areas in need of small business services (Small Business and Entrepreneurship Council 2021). Expanding small business ecosystems and ensuring ownership opportunities among historically excluded communities is essential to equitable green regeneration.

After decades of deindustrialization and deference to the private sector, smaller legacy cities finally have promising opportunities. The 2022 IRA aimed not only to invest significantly in climate-change issues but also to restore US manufacturing. For example, approximately $100 billion of the act's funding supports household electrification, from heat pumps and energy efficiency upgrades to electric vehicles (EVs) and charging stations (White House 2023). Significant federal tax credits and grants are also available to American manufacturers of low-carbon products, with requirements and incentives for use of parts sourced within the United States (Fischer 2022; Nahm, Lewis, and Allan 2022). Local government officials, accustomed to more traditional economic development practices like negotiating incentive packages, must also shift their approaches to leverage the massive opportunities ahead.

Local officials have good reason to nurture small businesses, because they keep wealth circulating within the community and expand the tax base. Both new small businesses and those poised to grow benefit from local government support far more than larger (usually corporate) entities, which have internal funding and in-house expertise. The 2008 Great Recession provided some insight into what modern small businesses need from local government and the ways economic development offices could effect low-carbon change (McFarland and McConnell 2013). They can

- conduct or commission baseline surveys to assess small business owners' needs, aspirations, and ideas about how the low-carbon transition might affect business plans;

- facilitate relationships with the state economic development office to help funnel federal funding;

- compile business and regulatory resources offered by the city and local and regional partners such as colleges, universities, and medical centers; financial institutions, including Community Development Financial Institutions (CDFIs), green banks, and credit unions; federal and state energy resources; and Workforce Investment Boards and chambers of commerce;

- help promote and coordinate financing to make small businesses energy efficient in communities of color (Kowalski 2021);

- coordinate collaboration among relevant local government offices to further small business growth, such as by zoning for walkable small business districts, better accommodating pedestrian traffic, or procuring locally owned goods and services; and

- maintain and periodically publicize a database of regional business incubator and mentorship programs, especially those geared to people of color and small businesses in low-income communities (Interise 2021).

CHAPTER 5

Building and Navigating
the Green Urban Policy Ecosystem

The 2022 Inflation Reduction Act and other federal actions have launched a slate of climate-related programs, including funding for electric vehicle charging infrastructure. *Source: Extreme Media/iStock/Getty Images Plus.*

Local leaders have a critical policy window for advancing the green regeneration of smaller legacy cities. A cohesive set of sustainability initiatives must connect, expand, and promote climate resilience, environmental justice and equity, and green economic development in smaller legacy cities. Creating such initiatives will enable a holistic suite of green policies and programs that maximize these cities' assets and surmount their underlying socioeconomic and environmental challenges. Several legacy cities have independent policies and programs that focus on one or two of these three priorities, but few—least of all smaller ones—have developed initiatives that achieve third-generation integration across all three policy areas.

To scale existing sustainability initiatives and build the requisite sustainability capacities, smaller legacy city officials must

- chart and assess the sustainability policy ecosystems in which their community operates;
- seek a diverse portfolio of resources, grants, and funding opportunities; and
- establish collaborative, cross-sector partnerships with a range of nonprofits, institutions, and intermediaries that can complement and expand municipal capacities.

Smaller legacy cities operate within a dynamic policy ecosystem comprising a range of supporters and/or opponents of green initiatives. The ecosystem map in figure 5.1 offers policymakers, practitioners, and partners a practical guide to aligning three pressing policy

areas—climate resilience, environmental justice and equity, and green economic development—within the priorities of each legacy city. Though by no means comprehensive, the map can also serve as a diagnostic tool to identify existing strengths, weaknesses, opportunities, and threats alongside more common external characteristics.

Local urban sustainability initiatives serve as the center or hub for this ecosystem as they oversee many of the policy and program levers related to their green initiatives. Within the core circle local governments experience internal forces, such as fiscal stability, organizational structure, and technical capacity, that can greatly influence the pace and trajectory of green policy changes in smaller legacy cities. The four outside circles represent common external drivers that impact smaller legacy cities:

Figure 5.1

Policy Ecosystem Map

(Schilling and Velasco 2021)

Economic and Market Forces. Longstanding economic decline and market instability found in many legacy cities—such as job and population loss, high poverty, fiscal instability, along with aging housing stock and infrastructure, and deteriorating neighborhood conditions—influence the capacity for interventions that local governments and their partners can take.

Federal and State Regulations and Resources. As we discuss throughout this report, federal and state green policies and programs, such as the 2022 Inflation Reduction Act or New York's 2019 Climate Initiative, are already having significant impacts on the scope, pace, and timing of legacy cities' efforts to advance sustainability.

Regional and Local Plans, Policies, and Programs. Often based on federal and state directives, regional agencies and local governments adopt and implement a suite of initiatives covering anything from transportation, economic development, and housing to water infrastructure, environmental protection, land conservation, and now climate resilience and equity.

Politics. These days federal, state, and local politics constantly reshape the contours of urban sustainability initiatives within smaller legacy cities as many public officials, and segments of the community at large, resist moving away from fossil fuels and fail to acknowledge climate change and the necessity of racial reckoning and equity.

Collectively, these drivers impact the existing and future capacities of smaller legacy city sustainability initiatives at every level.

Within their ecosystem, smaller legacy cities will find a range of cross-sector supporters (or opponents) of urban sustainability initiatives, though each community must also navigate uncertainty and customize coalitions, frameworks, and practices to ensure sustainability priorities align with local needs. Typical partners might include nonprofit and civic

organizations such as green community development corporations (CDCs), environmental justice advocates, and regional consortiums. Many business alliances, industry groups, and universities have sustainability initiatives and research programs, and international networks may also be useful resources.

As with any fundamental change, too, the pace for launching and institutionalizing new policies and programs across sectors unfolds in different stages and depends on many variables and actors. The receptivity of local leaders and the community at large can also shift over time. Crafting a narrative that describes the local impacts of climate change and the economic and equitable benefits of a low-carbon future can thus help make climate change more tangible and accessible to the community. As discussed in chapter 6, enlisting the support of green intermediaries and building capacity through a local office of sustainability is critical to developing and sharing that narrative.

Local governments cannot always directly access policy levers at the right scale for mitigating and adapting to climate change, so smaller legacy cities must take advantage of strategic and practical opportunities to prepare for historic transitions.

This dynamic, multilevel, high-stakes policy framework is challenging for any city, but with extremely limited resources and staff, policymakers in smaller legacy cities may find it especially hard to design, launch, and sustain their own climate resilience and equity programs. Local governments cannot always directly access policy levers at the right scale for mitigating and adapting to climate change, so smaller legacy cities must take advantage of strategic collaborations and practical opportunities to prepare for historic transitions.

Federal Sustainability and Climate Policy Frameworks

The federal government's climate actions ebb and flow with each administration. President Bill Clinton put sustainability on the federal policy agenda for the first time in 1993 when he convened the President's Council on Sustainable Development (Dernbach 2008; White House n.d.). Nearly 15 years later, the Obama Administration elevated climate mitigation and adaptation through numerous initiatives and new rules, including more stringent air-quality requirements for cars and power plants and interagency coordination through the Partnership for Sustainable Communities and the National Disaster Resilience Competition (Executive Office of the President 2013; US Department of Housing and Urban Development 2013; US EPA 2010b). Several of these efforts, including the Partnership for Sustainable Communities, provided direct technical assistance and capacity building that led to the adoption and implementation of dozens of regional and local sustainability plans and projects in cities, small towns, and rural areas across the country (Herberle et al. 2017).

While the Trump Administration later moved to repeal many of these preexisting climate policies, such as the power plant rule, and withdrew the country from the Paris Climate Accord, the Biden Administration has since prioritized climate change, environmental justice, and economic recovery. Within days of taking office, President Joe Biden issued the Justice40 executive order, requiring that 40 percent of designated federal investment benefits flow to disadvantaged communities that are marginalized, underserved, and overburdened by pollution. The order covers federal grants and programs that focus on climate change, clean energy, energy efficiency, clean transit, affordable and sustainable housing, training and workforce development, remediation and reduction of legacy pollution, and clean water and wastewater infrastructure. The White House Council on Environmental Quality, in collaboration with the Office of Management and Budget (OMB), has since developed the Climate and Economic Justice Screening Tool to help federal agencies identify designated disadvantaged communities eligible for support and provided additional guidance for federal agencies on implementation (Justice40 initiative 2022).

In Youngstown, Ohio, solar panels supply low-cost energy to the urban farm and kitchen incubator supported by the Youngstown Neighborhood Development Corporation. *Source: JM Schilling.*

New federal resources can help many smaller legacy cities replace and upgrade their decaying water infrastructure, as in Flint, Michigan, where such work nears completion. *Source: Jim West/Alamy Stock Photo.*

The Justice40 goals also apply to many initiatives outlined in two of the most significant federal legislative actions to address climate change within a generation—the Infrastructure Investment and Jobs Act (IIJA) and the Inflation Reduction Act (IRA). Each law establishes a slate of climate resilience, environmental justice, and GED grants, resources, and programs, such as

- climate and environmental justice block grants;
- transportation access and equity grants;
- pollution reduction and cleanup at ports;
- energy and water efficiency and climate resilience for affordable public housing;
- community urban forestry and tree planting;
- replacement of drinking water service lines;
- expansion of funds for superfund and brownfields remediation and reuse;
- new programs to modernize transit with clean technologies; and
- electric vehicle infrastructure.

Beyond the White House and its host of task forces and interagency working groups, the core federal agencies that steward the current climate resilience and equity agenda include the EPA, Department of Energy (DOE), the Department of Transportation (DOT), the Department of Housing and Urban Development (HUD), the Economic Development Administration (EDA), the Department of Agriculture (USDA), the Federal Emergency Management Agency (FEMA), and the National Oceanic and Atmospheric Administration (NOAA). Each has various regulatory roles and can take different actions to help states, local governments, and NGOs address climate resilience, environmental injustice, and GED. Each may also have special offices with targeted initiatives that provide capacity building and technical assistance to local governments and communities on environmental justice, equitable development, and climate resilience. For example, the Biden Administration's interagency Thriving Communities Network, housed at the Department of Transportation, coordinates the technical assistance programs of seven federal agencies (US Department of Transportation 2023).

Given the complexities of the policy ecosystem, the regional offices of federal agencies are a good place for smaller legacy cities to start exploring federal programs, as staff there typically work directly with regional communities on relevant programs and resources. Another place to seek assistance is with state environmental agencies, which have roles similar to those of their federal counterparts and are responsible for implementing federal policies and for allocating and overseeing federal grants to local governments and communities.

State Climate Equity Policy Frameworks

Several states, including two discussed below, have launched climate resilience initiatives, climate justice strategies, and other plans with a range of actions to move their economies toward clean energy and green jobs (Curry 2020). State-level climate initiatives typically focus on mitigation: setting emission targets, establishing carbon-pricing schemes, reforming the electricity power sector, expanding energy efficiency, transitioning to low-carbon transportation, and adopting state CAPs (Climate XChange 2023). Decarbonization—inventorying GHG emissions and reducing carbon pollution—has been those states' primary policy objective, with a goal of achieving a 100 percent clean-energy economy (Ricketts et al. 2020). As climate change has become an immediate threat in recent years, upgrading adaptation and resilience measures, with stress on vulnerability assessments and disaster planning, has also become a shared priority.

State climate initiatives can start with executive action by the governor, such as creating a CAP, and then supporting legislation, depending on the state's political dynamics. CAPs remain the most popular planning device, with 33 state CAPs released or in revision at the time of publication (Center for Climate and Energy Solutions 2022). Some states have comprehensive CAPs, along with more specific mitigation policies and programs that include a blend of strategic targets, regulatory actions, and resources for pilot projects—and many of these state efforts focus on transitioning private energy sectors and markets (Davis 2021). For the most part, however, states have separate initiatives for resilience strategies, administered by separate agencies, and only a few existing state climate initiatives include explicit racial equity programming—a gap that smaller legacy city policymakers should take special care to fill.

New York Climate Leadership and Community Protection Act (2019)

New York State contains a significant number of smaller legacy cities, and its climate initiative takes a holistic policy approach to carbon mitigation by focusing on three primary energy sectors—the electricity power grid, energy-efficient homes and businesses, and clean transportation—with a social-justice component to ensure an equitable, community-centric socioeconomic greening process (Center for Climate and Energy Solutions 2022). The state's Climate Leadership and Community Protection Act establishes a goal of 40 percent carbon emissions reduction by 2030 and 85 percent by 2050, with 100 percent of the state's electricity coming from clean energy sources at that point. It further requires the adoption of guidance and regulations that can foster decarbonization through buildings, transportation networks, and energy systems. The chartered Climate Action Council's scoping plan will further inform activities and help analyze regulatory changes; the council also includes the Just Transition Working Group and Climate Justice Working Group, to prioritize equity during this critical energy and market transformation (French 2022; New York State Climate Action Council 2022).

Massachusetts Clean Energy and Climate Plan for 2050 (2022)

The Massachusetts Clean Energy and Climate Plan (CECP) for 2050 is exceptionally nimble: one of its central features is that it can accommodate unforeseen economic, technological, policy, and natural developments. The latest in a series of plans issued by the state Executive Office of Energy and Environmental Affairs, CECP 2050 lays out near-term emission targets and "sector-specific sublimits" across transportation, residential HVAC, commercial and industrial HVAC, electric power generation, natural gas service, and in a broad industrial category that includes solid waste disposal and gas leakage repairs. It also includes benchmarks for carbon capture through water conservation, tree cover, and natural and working lands, and for aggressive workforce development for a rapidly expanding clean energy economy. Unlike its predecessors, it also stresses environmental justice remedies and benchmarks with stronger community engagement—including the availability of a statewide data dashboard—in working toward a statewide net-zero emissions goal (or 85 percent of 1990 GHG levels) by 2050. (Massachusetts Executive Office of Energy and Environmental Affairs 2022).

These state-level climate policy frameworks typically include a blend of incentives and rules ("carrots and sticks") to encourage private sector investment and actions. Along those lines, several state climate acts also include a handful of incentives for local government's energy, transportation, and green infrastructure projects. As more states enact similar climate mitigation frameworks, legacy city leaders should weigh in on their design to identify and carve out complementary roles, responsibilities, and resources for local governments.

Local legacy city leaders in and out of government with similar state climate initiatives should identify the range of actions that could make important contributions to their states' decarbonization targets and green economic development and environmental justice goals. Common actions might involve

- reducing the carbon footprint of city facilities, buildings, and operations;

- pursuing high-impact local policy opportunities (such as improving energy efficiency in homes and businesses, greening transportation systems, and streamlining or revising land use policies to facilitate the siting of renewable energy systems) to support the economic and small business transformations in many state climate plans;

- promoting state climate policies and supporting private sector efforts to adopt and implement state climate targets;

- protecting working agricultural land and developing local food systems; and

- weaving environmental justice and equity throughout the local government's climate resilience programming.

Intergovernmental Dynamics: Federal Regulation and State Preemption vs. Local Government Independence

Within the sustainability field, state agencies and powers can preempt local government policies and laws, especially regarding environmental regulations, solid waste management and recycling, and energy generation. State policies can also require or incentivize local development initiatives. Massachusetts, for example, subsidizes development in locally designated Smart Growth districts, with zoning ordinances requiring dense residential or mixed-use construction with a high percentage of affordable housing units (Commonwealth of Massachusetts 2017). California's Transformative Climate Communities program offers grants and technical assistance to help struggling cities and neighborhoods impacted by pollution to reduce GHGs through neighborhood-scale community development projects that foster environmental justice. (State of California n.d.).

Federal government agencies and regulations also play important roles in environmental, housing, and transportation infrastructure programs and policies. Sometimes local governments benefit from federal resources and agency actions, such as Brownfields Redevelopment grants or HUD's 2011–2015 Sustainable Communities Initiative's planning and challenge grants. But local governments with limited resources may still find compliance with federal and state rules burdensome, and many legacy cities are the subject of environmental litigation such as consent decrees that require costly but necessary infrastructure upgrades to their antiquated stormwater and sewer systems.

California's SB 375, the Sustainable Communities and Climate Protection Act of 2008, establishes a statewide network of regional plans with targets to reduce GHG emissions that focus particularly on vehicles. *Source: Remi Jouan/Wikimedia Commons.*

Regional Climate Resilience Policy Frameworks

During the 1990s and early 2000s, before many US federal and state government agencies became involved, local governments actively led climate mitigation through networks and associations such as the United States Conference of Mayors ("the Conference") and the international ICLEI—Local Governments for Sustainability. Cities and towns continue to learn from each other and share ideas through professional networks of mayors, local elected officials, and city and county managers (see table 2.1 on page 20 for a general list).

Yet many of the governance and infrastructure challenges, policy dimensions, and physical threats associated with climate change extend beyond single municipalities. Action at the metropolitan or even regional level is thus necessary for tackling climate mitigation and resilience, for equitable adaptation strategies, and even for meeting state and federal climate goals.

Despite the policy merits, however, regional climate resilience plans and initiatives are more prevalent in large, urban metropolitan areas—including Boston, Massachusetts; Miami, Florida; Los Angeles, California; Kansas City, Missouri; Minneapolis and St. Paul, Minnesota; and Washington, DC—partly because their regional consortiums and councils of government have more capacity and resources. Plus, successful regional planning of any kind is hard to pull off, much less to implement and enforce. Regional policy dynamics often include urban-suburban conflict, driven in part by a legacy of racial tension. Communities in smaller metropolitan regions are also accustomed to competing fiercely with one another for business, fueled by highly political exurban development pressures and tax burdens (Shi 2019).

State and federal executive authorities can require and fund some climate-directed regional actions in concert with regional transportation planning, economic and workforce development, and other forms of infrastructure. California's enactment of SB 375 in 2008 launched the country's first regional climate action regulations requiring local land use plans to comply with regional emission and carbon reduction goals set by the state (TransForm n.d.). However, most legacy cities have long been tied to the auto industry, and their built forms and cultures depend heavily on cars, leaving them less receptive to developing the transportation and land use alternatives on which climate action depends.

Although the concept of formal regional planning has existed for about a century, regional sustainability planning came of age with HUD's Sustainable Communities Initiative (SCI) during the Obama

Administration. That initiative awarded grants to 74 regions in 44 states over four years to design or implement regional sustainability plans (Mattiuzzi and Chapple 2020). With climate change again gaining policy traction and expanding federal funding during the Biden Administration, regional actors can develop or upgrade the climate and equity elements in their plans. Nonprofit intermediaries can also play a strong convening role in regional climate planning, as is happening in the smaller legacy cities in Northeast Ohio and in New York's Hudson Valley.

Part of the ReImagine Appalachia program focuses on green economic development strategies that would help industries and communities transition away from fossil fuels like coal, seen here on the barge approaching the Big Four Bridge in Louisville, Kentucky. *Source: Jeff Young.*

The regional scale also serves as the "sweet spot" for GED policy in smaller legacy cities, where green regeneration depends on the alignment of export potential, consumer and production markets, labor-market skills, recreational services, and educational networks. Legacy cities, even smaller ones, typically have economic or workforce development offices, but policy controls and economic development program and resource coordination still tend to originate at the county or regional level (OECD 2006). Public agencies and actors, then, often work closely with local and regional business alliances, chambers of commerce, and vocational education institutions, and together they can help small legacy city leaders and their partners navigate the relationships and dynamics of their GED policy ecosystem. Effective intergovernmental collaboration is essential for accelerating regional marketplace sustainability and scaling it down to the local level in smaller legacy cities.

Mayoral compacts can cover an even larger regional scale, as with the 2008 Massachusetts Gateway Cities Compact, which led to the creation of a legislative caucus, think-tank research, and a policy advocacy program (*SouthCoast Today* 2009). Despite organized interventions in education, housing, and poverty reduction, these cities—especially those farthest from Boston—still struggle. Yet they are also much better positioned to reap the economic benefits of the unfolding low-carbon economy. More recently, eight smaller legacy city mayors across four states in coal country have lent organized support for the ReImagine Appalachia initiative (see text box) to promote green regeneration.

Expanding Capacity Through Collaboration with Green Intermediaries and Networks

Because most smaller legacy cities have limited capacities, tracking and applying for state, federal, and philanthropic support for green regeneration programming can be extremely difficult (Blumgart 2022). Fortunately, a universe of intermediary organizations, operating on multiple scales with varying program emphases, is available to collaborate with local governments to advance climate resilience, environmental justice and racial equity, and GED.

Green intermediaries have an extensive track record of helping to build and expand the sustainability capacities of local governments and their partners through peer learning, technical assistance, and

ReImagine Appalachia

In late 2020, a pair of aligned initiatives called on Appalachian leaders to switch from coal, steel, and gas industries to a low-carbon-based economy and green regeneration. Eight smaller legacy city mayors across the four Ohio River Valley states of Kentucky, Ohio, Pennsylvania, and West Virginia called for "A Marshall Plan for Middle America" in the *Washington Post*, while a coalition of economic, environmental, and community organizations from the same region established a related action plan (Peduto et al. 2022).

The coalition's program, ReImagine Appalachia, currently has over 100 endorsers of its evolving economic development blueprint, which calls for a combination of place-based strategies and public investments designed to guide the private sector toward regional green regeneration (ReImagine Appalachia 2021). In alignment with recent federal funding priorities, its aims include

- repurposing underused land, such as reclaimed mines and capped landfills, to support renewable energy-generating infrastructure;
- modernizing the electric grid and expanding broadband to reach low-density and low-income communities;
- building a sustainable transport system, including EVs and transit;
- making the region a union-based manufacturing hub for EVs and alternatives to single-use plastics;
- ensuring equitable job access for Black residents, Indigenous residents, women, low-wage workers, and those facing employment barriers due to incarceration or addiction; and
- promoting local ownership models and community oversight for state and federal investments.

Table 5.1

Types of Green Intermediaries

Green Intermediary	Scale	Approach
Community-based organizations (CBOs)	Neighborhood or city	The community roots of green CBOs and longstanding relationships with neighborhoods help engage residents, build civic capacity, and foster resident power and leadership around urban greening, water quality, health, food security, small business support, jobs, and environmental and economic justice.
National, state, or local nonprofits	Various	Strong community relationships, expertise on sustainability policy and advocacy, and program flexibility enable nongovernmental organizations to facilitate cross-sector collaborations and bridge diverse communities, constituents, residents, and professions.
Education, community, and healthcare institutions	City or region	Universities, colleges, community foundations, and healthcare institutions support or host sustainability, urban design, and policy initiatives, centers, and programs that engage in service learning, community engaged research, capacity development, policy analysis, and a wide range of green community planning and design projects.
Government and quasi-governmental entities	Various	Government agencies and quasi-governmental entities often have regulatory, planning, and grant responsibilities for cities and towns. They can also provide supplemental capacity to coordinate state, regional, and local government officials' efforts to act on common issues and to scale solutions.
Networks	Various	Intermediary organizations play important roles as conveners, facilitators, mentors, coaches, researchers, and technical experts. Depending on the policy issue, a network could simultaneously operate at national, state, regional, local, or community levels.
Professional local government associations	National and state or regional chapters	National associations of local leaders provide professional development, training, policy analysis, research, and advocacy on pressing public and community problems, and they represent local governments' interests in federal and state policy and legislation.

collective-impact policymaking. They act as conveners, coordinators, funders, and technical assistance providers for smaller legacy cities that can help them connect the different dimensions of the green policy ecosystem (see figure 5.1). By devoting time and energy to partnering with green intermediaries amid climate transitions, smaller legacy cities can continue to evolve and become greener. A representative list of intermediaries, some of which work in and with legacy cities, appears in table 5.1.

Smaller legacy cities may not have sufficient capacity to undertake all their desired activities alone, so partnering with intermediaries and coordinating with federal, state, and regional agencies ensures local green regeneration actions have shared support—and funding. For example, in 2016 and 2017, Providence, Rhode Island, received $125,000 in grants through the Partners for Places Equity Pilot initiative, a joint project of the Funders' Network for Smart Growth and Livable Communities and

In practice	Greening-related examples
Some smaller legacy cities partner with green community development corporations (CDCs) that apply an environmental justice, climate resilience, and sustainability lens to their traditional work to revitalize neighborhoods, build affordable housing, cultivate community gardens, maintain parks and public spaces, and organize and empower local residents.	Cleveland Neighborhood Progress, PUSH Buffalo, Youngstown Neighborhood Development Corporation (YNDC), individual Groundwork USA communities
Many green intermediaries supplement and assist local governments in providing essential capacities such as neighborhood planning, data gathering, and nuisance abatement, along with technical assistance for community projects and green infrastructure.	Groundwork USA (national and local), Alliance for Sustainable Communities–Lehigh Valley (regional), Center State CEO (regional), Enterprise Green Communities (national), Ohio Climate Justice Fund (state)
Professors and students collaborate with local governments, nonprofit partners, and community members in developing innovative solutions, while community foundations provide resources and strategic guidance to support these centers, staff, and students.	University of Wisconsin–Madison's Center on Wisconsin Strategy, Penn State's Sustainable Communities Collaborative, SUNY College of Environmental Science and Forestry (various Syracuse environmental design programs)
Metropolitan planning organizations and regional councils of government may serve as hubs for capacity building and technical assistance to smaller local governments around environmental, transportation, economic, and planning challenges. Other agencies, such as land banks and redevelopment authorities, can apply their designated powers to acquire, manage, and redevelop underused or vacant properties.	Federal Reserve Bank of Boston Working Communities Challenge (regional), New Jersey Office of Environmental Justice, Cuyahoga Land Bank (regional), Mid-America Regional Council in Kansas City, Missouri (local and regional)
Intermediaries facilitate knowledge sharing and peer learning, advance common policy and political agendas, and undertake research to build support for local-level policy and program actions.	GARE: Government Alliance on Race and Equity (national), Gateway Cities Innovation Institute (state), Sustainable States Network (national), Future of Small Cities Institute (regional)
Associations or their chapters sponsor conferences, workshops, and communities of practice that engage members (often in cohorts), facilitate peer learning, and share innovative policies on sustainability, climate resilience, infrastructure, equitable development, racial justice, economic mobility, and more.	United States Conference of Mayors, Climate Mayors, National League of Cities, American Planning Association, International City/County Management Association, National Association of Counties, Bloomberg Harvard City Leadership Initiative, Urban Sustainability Directors Network

the Urban Sustainability Directors Network, that supported a partnership among the city's Office of Sustainability, the Environmental Justice League of Rhode Island, and Groundwork Rhode Island to embed equity goals, strategies, actions, and metrics into the citywide sustainability plan (City of Providence 2017a).

In many respects, green regeneration programming can serve as a natural bridge across intergovernmental policy silos and facilitate regional climate equity and resilience to connect multiple local governments, including smaller legacy cities, and the private and nonprofit sectors. Beyond better inter- and intra-governmental collaborations, smaller legacy cities can capitalize on their community assets—physical, civic, environmental, and economic—and build new partnerships with nearby cities and various green intermediaries for greater impact and shared efforts.

CHAPTER 6
Conclusions and Recommendations

The City of Detroit, Michigan, demolished abandoned and deteriorating Midtown homes to pave the way for neighborhood revitalization—a rite of passage for many evolving legacy cities. *Source: JM Schilling.*

Smaller legacy cities can be innovators in green regeneration, drawing on their regional land and water assets, surplus infrastructure, culture, and history. Yet the myriad threats of the global climate crisis, combined with the specific conditions these cities face—economic disintegration, depopulation, deep racial inequity, and environmental degradation—elude traditional "market-based solutions." Instead, smaller legacy cities require a new generation of policies and resources that respond to their socioeconomic and capacity constraints and that center equity and justice alongside economic development and sustainability.

The following recommendations provide a framework for smaller legacy city leaders—in and adjacent to government—seeking guidance on preparing for climate change, as well as for their regional, state, and federal allies and partners. Near- and long-term sustainability programming must at every level integrate climate resilience, environmental justice, and green economic development initiatives, and success depends on deft navigation of regional intergovernmental policy ecosystems. Strengthening smaller legacy city capacities through cross-sector partnerships with intermediary organizations is not only an achievable goal, it is also essential to mitigating GHG emissions and adapting to climate change across the United States.

LEVERAGE GREENER LAND USE PLANS AND CODES

Working closely with community participants, local officials should screen comprehensive plans and zoning ordinances for climate resilience and equity features, and inventory existing buildings, housing, and neighborhoods to better inform comprehensive and sustainability planning. Low-Impact Development (LID) interventions, for example, can be made while upgrading sewer infrastructure. They are less expensive than conventional pipe-based drainage and can often support jobs in landscaping and maintenance.

Blue-green infrastructure, including urban forestry programs, reduces stormwater overflow, pollution, flooding, and urban heat island effects while also providing effective carbon sinks. In addition, greening the abundant vacant lots in smaller legacy cities with parks, community gardens, recreational spaces, and side-lot purchases contributes to both climate resilience and quality of life. On a regional scale, multi-modal transportation planning and farmland protection can improve food production and availability, and watersheds and undeveloped land should be managed to enhance carbon sequestration, water filtration, and other ecosystem services.

EMBED ENVIRONMENTAL JUSTICE IN GREENING INTERVENTIONS

Environmental racism is embedded in longstanding urban land use practices that have historically excluded people of color from suburban development while exposing their city neighborhoods to harmful toxins, deterring homeownership and other forms of investment as well as degrading public health.

Beyond the competitive grants outlined in the IIJA and IRA, state and local officials should prioritize traditional statutory formula grants to ensure more equitably distributed and resilient infrastructure systems are developed to meet the demands of sustainability in the twenty-first century. Systems like water, energy, transportation, broadband, and other digital technologies should comply with the federal government's Justice40 goal of allotting 40 percent of these investments to neighborhoods with long histories of racial discrimination and disinvestment.

Working closely with community organizations with street-based contacts and expertise, cities must map out trouble spots and proactively mitigate further harm while also banning toxic sites—existing and under development—near densely populated neighborhoods. Zoning ordinances can identify where specific interventions should be prioritized for conditions such as deteriorated housing, lead paint and water contamination, transportation deficits, and poor sanitation.

Working closely with community organizations with street-based contacts and expertise, cities must map out trouble spots and proactively mitigate further harm while also banning toxic sites—existing and under development—near densely populated neighborhoods.

At the federal level, the EPA's long-running brownfield conversion work has broadened by incorporating new capacity-building grants to environmental justice intermediaries. Brownfields work is also moving from single-building projects to "area-wide" planning that covers multiple structures and land uses. The 2021 IIJA basically doubles brownfields funding for the next five years, and green intermediaries, such as the nonprofit Groundwork USA, have turned these projects into inclusive planning opportunities and job-training programs, with a focus on smaller legacy cities such as Lawrence, Massachusetts; Elizabeth, New Jersey; and Buffalo, New York.

DECARBONIZE ENERGY INFRASTRUCTURE AND ELEVATE ENERGY EQUITY

Smaller legacy cities contain and are usually surrounded by an abundance of underused land that can be repurposed for clean energy generation. Local governments should strategically license urban brownfields, dead malls and big-box stores, and capped mines and landfills to generate distributed or utility-scale renewable energy—and incentivize utilities to join production. In 22 states, offsite community solar subscription programs provide renters as well as homeowners the economic benefits of renewable energy, and more communities could easily get on board.

Although extremely difficult to achieve, municipal utility ownership is fiscally and operationally preferable to investor-owned utilities and therefore worth considering. In lieu of that, smaller legacy city leaders should familiarize themselves with state permitting, engineering, and private utility oversight processes to plan accordingly and attract private development. In states with the appropriate enabling statutes, advocates and legislators should implement community choice aggregation, which reduces costs and makes it easier for neighborhoods to choose renewable energy sources.

Electric grid improvements are crucial not only to the low-carbon electrification of HVAC and transportation systems but also to overcoming the digital divide in

Ithaca, New York, is implementing its new Efficiency Retrofitting and Thermal Load Electrification Program to achieve community-wide carbon neutrality by 2030, a commitment that puts it on the cutting edge of small-city decarbonization. *Source: TW Farlow/iStock/Getty Images Plus.*

The City of Troy, New York, launched a Sustainability Task Force in 2014 to determine how the city could address GHG emissions, the impacts of climate change, and greener city operations. *Source: ChrisBoswell/iStock/Getty Images Plus.*

the economic and educational opportunities available through broadband access. Aside from harnessing state funding, smaller legacy cities should ensure all residents have proper services, especially in vulnerable and rural communities, which requires both careful resident engagement and intentional shepherding of consumer subsidies to internet service providers (Mazella 2021).

Decarbonization efforts—as well as household, government, and business budgets—are poorly served if heating and cooling are not adequately insulated within a building, and landlords who do not cover energy bills are often indifferent to these upgrades. Through individual states, the DOE's Weatherization Assistance Program subsidizes low-income homeowners; local government outreach, coordinated with nonprofit advocates, can organize both tenants and landlords to find common ground—and even help landlords and homeowners identify related renovation-assistance programs, such as heat-pump furnace replacements and lead removal. Building codes should also require both new building and retrofits to meet stringent energy-efficient standards.

Because most climate-transition work will be in traditional trades—particularly construction, electricity, plumbing, and maintenance—local partners can work with unions, regional workforce development boards, and vocational technical programs to ensure that affordable training and high-quality career tracks are available to a diverse workforce in these fields (Gallucci 2019).

EMPOWER A STAND-ALONE SUSTAINABILITY OFFICE

Although many capacity-challenged smaller legacy cities have cobbled together first- or second-generation sustainability programming, most have not had the means or mandate to establish a permanent, full-fledged, well-staffed sustainability office whose mission is integrated throughout local government. Recent federal funding for water, transportation, grid infrastructure, and global climate action has changed the sustainability landscape dramatically, however, signaling that more private and philanthropic investment is on the way. Green regeneration is on the horizon, but it will require skilled staff to attract and manage funding, develop programming, and regularly engage residents, local and state officials, nonprofit intermediaries, and learning networks. Smaller legacy cities must organize to take full advantage of these funding opportunities while they last.

To that end, the ideal municipal sustainability department would have the ear of city leadership, as well as of state agencies that oversee the federal funds flowing to localities. With staff to track and manage funding, buoyed by midcareer, multiyear professional fellows, a director must elevate community voices, empower neighborhoods to tackle climate change on the front lines (especially in communities of color), and otherwise cultivate strong community advocacy.

A sustainability director can also help implement equitable and sustainable public procurement and operations rules at the local level. Procedural steps to green city car fleets, facilities, and buildings or to incentivize inclusive purchasing from local businesses and event planners can go a long way.

LAUNCH DEDICATED COMMUNITY ENGAGEMENT EFFORTS

Transitioning to an inclusive low-carbon economy is a hyperlocal effort that demands broad cooperation and civic accountability. Given the decline of local news outlets, local governments and their nonprofit partners should establish effective, active community engagement platforms and programs to keep the public informed and their voices heard.

Business and community members as well as local officials must participate in both formal and informal processes to share specialized knowledge, such as with recent federal funding opportunities in semiconductor manufacturing, infrastructure, and climate resilience. All these groups must find common ground focused on the importance of climate action and racial equity to economic revitalization. Local economic development officials can assess existing efforts, convene community stakeholders, and consult the equitable economic development playbook for successful strategies (International Economic Development Council 2023).

Local partners with standing in their communities can also advise on grassroots concerns and disseminate new information organically—though such relationships require trust and authenticity. Officials and advocates alike might consider establishing an environmental justice charter, opening more pathways to policy influence, supporting community organizations and the knowledge they bring to the table by ceding decision-making, providing financial compensation, and engaging in participatory budgeting.

Throughout community engagement efforts, partners should base greening messages on pressing local priorities such as generating jobs, improving health and health-care costs, lowering energy costs, and reducing flooding and urban heat. Overall, john a. powell's "targeted universalism" can be a helpful approach for committing the entire community to equitable development, with targeted strategies based on cultural and economic backgrounds (powell et al. 2019).

SUPPORT LOCAL SMALL BUSINESSES

Small businesses are crucial to keeping wealth generated by green economic development circulating within the community. Local governments should do everything they can to ensure new federal funding for equitable infrastructure, manufacturing, and clean energy goes to small businesses that support green job growth and career ladders, keep wealth circulating locally, and increase the tax base. Municipal programs should regularly communicate with state offices responsible for disseminating funds; they should also connect small businesses with local and regional institutional partners and other resources and integrate small business needs into all relevant municipal departments.

Local governments should do everything they can to ensure new federal funding for equitable infrastructure, manufacturing, and clean energy goes to small businesses that support green job growth and career ladders, keep wealth circulating locally, and increase the tax base.

Mentorship and business development programs, such as those offered by national nonprofit Interise, seek to fill knowledge and capital-access gaps among people of color in low-income communities. Local officials can nurture such efforts with publicity and funding, and help coordinate participation by local

institutions such as schools and colleges, banks and CDFIs, insurance companies, business organizations, and community-based nonprofits.

ACT REGIONALLY

Partly because of their size and partly due to their reduced budgets, smaller legacy cities simply have diminished capacity, technically and economically, making regional alliances crucial. When regions share ecological characteristics, economic history, transportation networks, and history, collaboration and burden sharing enable more effective and cost-efficient local climate resilience and green regeneration programming.

Multicity- or multistate-scale programs, such as Massachusetts Gateway Cities and ReImagine Appalachia, can be particularly effective for peer learning, lobbying state governments, launching larger community and economic development initiatives, and unifying smaller legacy cities in climate resilience. Interstate governance of joint economic endeavors like port authorities, resource management, or GED

efforts are also worth pursuing; examples include the nearly 70-year-old Great Lakes Commission and the newer 12-state Regional Greenhouse Gas Initiative (RGGI) to reduce carbon emissions in New England and the mid-Atlantic states.

BUILD CROSS-SECTOR CAPACITY, NETWORKS, AND POLICY ADVOCACY

Given their capacity challenges, smaller legacy city leaders and local partners simply need greening support. Legacy cities can begin by assessing the landscape of green intermediaries within their regions or states—such as colleges, nonprofit groups, climate equity advocates, green infrastructure, or sustainability alliances—that may be working in their own communities or nearby. They can also speak with community foundations and business leaders interested in sustainability and explore potential collaborations with national and statewide intermediaries that already have programming for environmental justice, climate resilience, and green economic development.

National or even statewide intermediaries, however, may not have much experience working with or in smaller legacy cities. Thus, these intermediaries, in collaboration with local leaders and their partners, must recalibrate and customize programming to ensure they offer the meaningful, tailored assistance smaller legacy cities require, such as grant writing, data systems management, civic engagement activities, and technical training.

Whether these national or local intermediaries are nonprofits, CBOs, government entities, or other local institutions, they too will need investments to expand and tailor their work. Ideally, state, regional, and local offices working with national and regional foundations would form a consortium to fund intermediaries that help officials and community actors with climate resilience planning and programming (Janus 2022).

Educational institutions are especially important allies. From public schools and community colleges to four-year colleges and universities, they can offer strategic training programs and help ensure equity in hiring, compensation, internships, and apprenticeships. Much of this work is regional and involves networking with nearby communities on workforce investment boards and state regional councils, and with trade associations and union locals.

The brutal effects of climate change are hardly restricted to city lines. Smaller legacy city leaders must embrace innovative technological, financing, and planning ideas—not only from their peers but also from those working in better-resourced small cities and larger legacy cities. National and regional organizations and alliances of local officials—such as Climate Mayors, the National League of Cities, and the Gateway Cities Innovation Institute in Massachusetts—offer effective networking and peer learning opportunities. The International City/County Management Association, state CBO umbrella groups, trade organizations and unions, the Urban Sustainability Directors Network, the Climate Justice Alliance, and other organizations offer new ideas for the policy pipeline.

Climate advocates in Pittsburgh, Pennsylvania, marched in 2021 to call for a just and intersectional transition to renewable energy in their large legacy city. *Source: Mark Dixon/Creative Commons.*

References

Adaptation Clearinghouse. 2015. "Growing Stronger: Toward a Climate-Ready Philadelphia." https://www.adaptationclearinghouse.org/resources/growing-stronger-toward-a-climate-ready-philadelphia.html.

Agyeman, Julian, David Schlosberg, Luke Craven, and Caitlin Matthews. 2016. "Trends and Directions in Environmental Justice: From Inequity to Everyday Life, Community, and Just Sustainabilities." *Annual Review of Environment and Resources* 41(1): 321–40. https://doi.org/10.1146/annurev-environ-110615-090052.

American Planning Association. 2021. "Daniel Burnham Award for a Comprehensive Plan: Richmond 300: A Guide for Growth." https://planning.org/awards/2021/excellence/richmond-300.

American Lung Association. 2022. "Disparities in the Impact of Air Pollution." (November). https://www.lung.org/clean-air/outdoors/who-is-at-risk/disparities.

American Public Health Association. 2019. "Addressing Environmental Justice to Achieve Health Equity." (November 5). https://www.apha.org/policies-and-advocacy/public-health-policy-statements/policy-database/2020/01/14/addressing-environmental-justice-to-achieve-health-equity.

Argerious, Natalie Bicknell. 2022. "Urbanism 101: What Is an EcoDistrict?" *The Urbanist*. (July 2). https://www.theurbanist.org/2022/07/02/urbanism-101-what-is-an-ecodistrict/.

Armstrong, Michael, Derik Broekhoff, Katherine Gajewski, Miya Kitahara, Michael McCormick, Sarah McKinstry-Wu, Ariella Maron, Hoi-Fei Mok, Tracy Morgenstern, Michael Steinhoff, and Brian Swett. 2021. "The State of Local Climate Planning: Observations by Local Climate Action Practitioners." CityScale. https://cityscale.org/wp-content/uploads/2021/05/State-of-Local-Climate-Planning.pdf.

Asimow, Noah, and Brooke Kushwaha. 2022. "Offshore Wind Projects Jockey for Position in Vineyard Waters." *Vineyard Gazette*. (November 3). https://vineyardgazette.com/news/2022/11/03/offshore-wind-projects-jockey-position-vineyard-waters.

Barton & Loguidice, DPC. 2016. "Green Infrastructure Retrofit Manual." City of Rochester and Monroe County. https://seagrant.sunysb.edu/coastalcomm/pdfs/GreenInfrastructure-Manual.pdf.

Bloomberg Cities. 2020. "5 Things to Consider When Doing a City Data Inventory." *Medium* (blog). (February 26). https://bloombergcities.medium.com/5-things-to-consider-when-doing-a-city-data-inventory-dbeeb9a6780c.

Blumgart, Jake. 2022. "Smaller Cities, Poorer Regions Could Be Infrastructure Losers." (January 11). *Governing*. https://www.governing.com/now/smaller-cities-poorer-regions-could-be-infrastructure-losers.

Bohnenberger, Katharina. 2022. "Greening Work: Labor Market Policies for the Environment." *Empirica* 49 (February 3): 347–368. https://doi.org/10.1007/s10663-021-09530-9.

Brinkley, Douglas. 2022. *Silent Spring Revolution: John F. Kennedy, Rachel Carson, Lyndon Johnson, Richard Nixon, and the Great Environmental Awakening*. New York, NY: Harper Collins.

Brookstein, Pamela, and Elena Savona. 2019. "Training Real Estate Professionals to Find the Value of Solar." Technical report for Elevate Energy. https://doi.org/10.2172/1532752.

Brown, Marcia. 2015. "When a Vacant Lot Goes Green: Best Practices After Demolition." *Policy Matters Ohio*. (September). https://www.policymattersohio.org/wp-content/uploads/2015/09/Blight-final1.pdf.

Bullard, Robert, ed. 1993. *Confronting Environmental Racism: Voices from the Grassroots*. Boston, MA: South End Press.

California Office of Environmental Health Hazard Assessment. n.d. "CalEnviroScreen." https://oehha.ca.gov/calenviroscreen.

Center for Climate and Energy Solutions. 2019. "What Is Climate Resilience and Why Does It Matter?" https://www.c2es.org/document/what-is-climate-resilience-and-why-does-it-matter.

Center for Climate and Energy Solutions. 2022. "US State Climate Action Plans." Map. (December). https://www.c2es.org/document/climate-action-plans.

Center for Community Progress. 2022. "National Land Bank Map." (March). https://communityprogress.org/resources/land-banks/national-land-bank-map.

Change Lab Solutions. n.d. "Complete Streets Policies at the Local Level." *Change Lab Solutions*. https://www.changelabsolutions.org/product/complete-streets-policies-local-level.

City of Philadelphia, PA, Office of Sustainability. 2016. "Greenworks: A Vision for a Sustainable Philadelphia." https://www.phila.gov/documents/greenworks-a-vision-for-a-sustainable-philadelphia.

City of Providence, RI, Office of Sustainability. 2017. "Equity in Sustainability: A Collaborative Initiative by the City of Providence and Frontline Communities of Color of Providence to Bring a Racial Equity Lens to the City's Sustainability Agenda." https://www.providenceri.gov/wp-content/uploads/2017/02/Equity-and-Sustainability-SummaryReport-2-20-reduced.pdf.

———. 2017. "City of Providence Awarded $125,000 Equity in Sustainability Grant." Press release. (May 5). https://www.providenceri.gov/city-providence-awarded-125000-equity-sustainability-grant.

———. 2019. City of Providence, *Climate Justice Plan*. https://www.providenceri.gov/sustainability/climate-justice-action-plan-providence.

City of Richmond, VA. 2020. "Richmond 300: A Guide for Growth." https://www.rva.gov/planning-development-review/master-plan.

City of Springfield, MA. 2015. National Disaster Resilience Competition, Phase II. (October 27). https://www.springfield-ma.gov/planning/fileadmin/community_dev/DR/NDRC_Phase_II_Complete_Application_public.pdf.

City of Springfield, MA, Office of Community Development. 2017. "Strong, Healthy & Just: Springfield's Climate Action & Resilience Plan." (June). https://www.pvpc.org/projects/strong-healthy-just-springfield-climate-action-resilience-plan.

City of Springfield, MA. 2021. *Community Action Plan*. https://www.springfield-ma.gov/comm-dev/fileadmin/community_dev/Action_Plan_2021-2022/Final_AP.pdf.

Cleary, Kathryne, and Karen Palmer. 2022. "Renewables 101: Integrating Renewable Energy Resources into the Grid." Resources for the Future. (August 15). https://www.rff.org/publications/explainers/renewables-101-integrating-renewables.

Cleveland Urban Design Collaborative. 2009. "Re-Imagining Cleveland: Vacant Land Re-Use Pattern Book." https://www.birminghamal.gov/wp-content/uploads/2017/08/6_Cleveland-vacant-land-pattern-book.pdf.

Climate Change Resource Center. n.d. "i-Tree." US Department of Agriculture. https://www.fs.usda.gov/ccrc/tool/i-tree.

Climate XChange. 2023. "The State Climate Policy Network." https://climate-xchange.org/network.

Climigration Network. n.d. "Climigration Network." https://www.climigration.org.

Coalition for Urban Transitions. n.d. "Seizing the Urban Opportunity." https://urbantransitions.global/urban-opportunity/seizing-the-urban-opportunity.

Cole, Luke, and Sheila Foster. 2000. *From the Ground Up: Environmental Racism and the Rise of the Environmental Justice Movement*. New York, NY: New York University Press. https://nyupress.org/9780814715376/from-the-ground-up.

Coming Clean, Inc. 2021. "The Louisville Charter for Safer Chemicals." https://comingcleaninc.org/louisville-charter/endorse.

Commonwealth of Massachusetts. 2017. "Chapter 40R." Housing and Community Development. (December 29). https://www.mass.gov/service-details/chapter-40r.

Commonwealth of Massachusetts. n.d. "Greening the Gateway Cities Program." https://www.mass.gov/service-details/greening-the-gateway-cities-program.

Connolly, James J., Dagney G. Faulk, and Emily J. Wornell. 2022. *Vulnerable Communities, Research, Policy and Practice in Small Cities*. Ithaca, NY: Cornell University Press.

Cowell, Margaret M. 2013. "Bounce Back or Move on: Regional Resilience and Economic Development Planning." *Cities* 30 (February): 212–22. https://doi.org/10.1016/j.cities.2012.04.001.

Crenshaw, Kimberlé Williams. 2020. "The Unmattering of Black Lives." *The New Republic* (May 21). https://newrepublic.com/article/157769/unmattering-black-lives.

Curry, Melanie. 2020. "What Does It Take to Meet State Climate and Equity Goals?" *Streetsblog California* (blog). (November 6). https://cal.streetsblog.org/2020/11/06/what-does-it-take-to-meet-state-climate-and-equity-goals.

Curtis, E. Mark, and Ioana Marinescu. 2022. "Green Energy Jobs in the US: What Are They, and Where Are They?" Working paper No. 30332. Cambridge, MA: National Bureau of Economic Research (August). https://www.nber.org/papers/w30332.

Davey Resource Group, Inc. 2018. "Tree Management Plan: City of Binghamton, New York." (October). https://www.binghamton-ny.gov/home/showpublisheddocument/1110/637582669266330000.

Davis, Eric. 2021. "How Do States Plan to Meet Their Climate Commitments?" Climate XChange. (July 14). https://climate-xchange.org/2021/07/14/how-do-states-plan-to-meet-their-climate-commitments/.

De Guire, Jeannette. 2012. "The Cincinnati Environmental Justice Ordinance: Proposing a New Model for Environmental Justice Regulations by the States." *Cleveland State Law Review* 60(1).

Dennis, Brady. 2022. "The Toll Extreme Weather Took in the US During 2022, by the Numbers." *Washington Post*. (December 30). https://www.washingtonpost.com/climate-environment/2022/12/30/blizzard-hurricane-drought-flood-tornado-2022.

Dennis, Brady, and Sarah Kaplan. 2021. "Humans Have Pushed the Climate into 'Unprecedented' Territory, Landmark U.N. Report Finds." *Washington Post*. (August 10). https://www.washingtonpost.com/climate-environment/2021/08/09/ipcc-climate-report-global-warming-greenhouse-gas-effect/.

Dernbach, John C. 2008. "Learning from the President's Council on Sustainable Development: The Need for a Real National Strategy." *Environmental Law Reporter* 32: 10648–10666. (February).

DiNapoli, Thomas P. 2022. "Green and Growing: Employment Opportunities in New York's Sustainable Economy." (February). Office of the New York State Comptroller. https://sps.columbia.edu/news/green-jobs-and-transition-environmentally-sustainable-economy.

Duchene, Courtney. 2022. "Philly's Climate Change Report Card." (June 21). Philadelphia Citizen. https://thephiladelphiacitizen.org/philly-climate-report-card.

Elam, Lindsey. 2019. "Downtown Newark Revitalization." *Greater Ohio Policy Center*. (October 7). https://www.greaterohio.org/good-ideas/2019/10/1/downtown-newark-revitalization.

Elevate Energy. 2023. "Research & Innovation." https://www.elevatenp.org/research-and-innovation.

Eley, Carlton. 2010. "Equitable Development: Untangling the Web of Urban Development Through Collaborative Problem Solving." *Sustain* 21 (March 29): 3–12. https://www.slideshare.net/celey/equitable-development-untangling-the-web-of-urban-development-through-collaborative-problem-solving.

Energy News Network. 2018. https://energynews.us/2018/04/18/solar-installations-put-ypsilanti-michigan-on-the-map-for-clean-energy.

Environmental Advocates of New York. 2019. "Climate Leadership and Community Protection Act: An Overview." https://eany.org/wp-content/uploads/2019/10/clcpa_fact_sheet_0.pdf.

Environment America. 2019. https://environmentamerica.org/michigan/center/media-center/ypsilanti-ranked-as-solar-leader-in-national-report.

Executive Office of the President. 2013. "The President's Climate Action Plan." https://obamawhitehouse.archives.gov/energy/climate-change.

Feiden, Wayne M., and Elisabeth Hamin. 2011. *Assessing Sustainability: A Guide for Local Governments*. American Planning Association, 112.

First National People of Color Environmental Leadership Summit. 1991. "Principles of Environmental Justice." https://www.ejnet.org/ej/principles.html.

Fischer, Anne. 2022. "Clean Energy Manufacturing Support in Inflation Reduction Act." *pv Magazine USA*. (July 29). https://pv-magazine-usa.com/2022/07/29/clean-energy-manufacturing-support-in-inflation-reduction-act.

Fitzgerald, Joan. (2010). *Emerald Cities: Urban Sustainability and Economic Development*. New York: Oxford University Press.

Fowler, Holly, and Kate O'Brien. 2017. "Planning with an Eye Toward Implementation: What All Communities Can Learn from Using a Brownfields Area-Wide Planning Approach." *Groundwork USA* 36. (December). https://groundworkusa.org/wp-content/uploads/2017/12/GWUSA_Planning-with-an-Eye-Toward-Implementation_BFAWP_Report_final.pdf.

Freitag, Robert C., Daniel B. Abramson, Manish Chalana, and Maximilian Dixon. 2014. "Whole Community Resilience: An Asset-Based Approach to Enhancing Adaptive Capacity Before a Disruption." *Journal of the American Planning Association* 80(4): 324–35. https://doi.org/10.1080/01944363.2014.990480.

French, Marie J. 2022. "New York Passes Sweeping Plan to Reduce Emissions and 'Lead the Way on Solving Climate Change.'" *Politico*. (December 19). https://www.politico.com/news/2022/12/19/new-york-emissions-climate-change-00074600.

Froias, Steven. 2022. "Connecting to the Offshore Wind Industry in New Bedford." *NBEDC News*. June 24. https://nbedc.org/connecting-to-the-offshore-wind-industry-in-new-bedford.

Futrell, Janae. 2019. "How to Design Your Scenario Planning Process." American Institute of Certified Planners. PAS memo. (July–August). https://www.planning.org/pas/memo/2019/jul/.

Future of Small Cities Institute. n.d. "The Future of Small Cities Institute." https://www.futureofsmallcities.org.

Gallucci, Maria. 2019. "Energy Equity: Bringing Solar Power to Low-Income Communities." *Yale E360*. (April). https://e360.yale.edu/features/energy-equity-bringing-solar-power-to-low-income-communities.

Gauna, Eileen. 2015. "Failed Promises: Federal Environmental Justice Policy in Permitting." In *Failed Promises: Evaluating the Federal Government's Response to Environmental Justice*, ed. David M. Konisky. Cambridge, MA: MIT Press. https://doi.org/10.7551/mitpress/9780262028837.003.0003.

Geertsma, Meleah. 2019. "Transforming Local Policies to Achieve Environmental Justice." Natural Resources Defense Council. (February 22). https://www.nrdc.org/resources/transforming-local-policies-achieve-environmental-justice.

Georgetown Climate Center. n.d. "Equitable Adaptation Legal & Policy Toolkit: Participatory Budgeting." https://www.georgetownclimate.org/adaptation/toolkits/equitable-adaptation-toolkit/participatory-budgeting.html.

Glaeser, Edward. 2018. "Mission: Revive the Rust Belt." *City Journal*. (October). https://www.city-journal.org/revive-rust-belt.

Goldstein, Benjamin, Dimitrios Gounaridis, and Joshua P. Newell. 2020. "The Carbon Footprint of Household Energy Use in the United States." *PNAS* 117(32): 19122–19130. https://www.pnas.org/doi/10.1073/pnas.1922205117.

González, Rose, founder of Facilitating Power. 2019. "The Spectrum of Community Engagement to Ownership." Movement Strategy Center. https://movementstrategy.org/resources/the-spectrum-of-community-engagement-to-ownership.

Greater Toledo Community Foundation. 2021. "Solar Field Generates Power, Revenue and Hope." *Toledo Blade*. (March 15). https://www.toledoblade.com/b-partners/gtcf/2021/03/15/Solar-field-generates-power-revenue-and-hope/stories/20210315001.

Groundwork Ohio River Valley. n.d. "Native Plant Propagation." https://www.groundworkorv.org/native-plant-propagation.

Harvey, Fiona. 2023. "Scientists Deliver 'Final Warning' on Climate Crisis: Act Now or It's Too Late." *The Guardian*. March 20. https://www.theguardian.com/environment/2023/mar/20/ipcc-climate-crisis-report-delivers-final-warning-on-15c.

Heckert, Megan, Joseph Schilling, and Fanny Carlet. 2016. "Brief No. 1: Greening Legacy Cities—Vacant Property Research Network." Vacant Property Research Network. https://vacantpropertyresearch.com/translation-briefs/greening.

Herberle, Lauren C., Brandon McReynolds, Steve Sizemore, and Joseph Schilling. 2017. "HUD's Sustainable Communities Initiative: An Emerging Model of Place-Based Federal Policy and Collaborative Capacity Building." *Cityscape* 19(3).

Hoffman, Jeremy S., and Rachel Dunn. 2021. "Where Do We Need Shade? Mapping Urban Heat Islands in Richmond, Virginia." US Climate Resilience Toolkit. (June 4). https://toolkit.climate.gov/case-studies/where-do-we-need-shade-mapping-urban-heat-islands-richmond-virginia.

Hollingsworth, Torey. 2016. "From Akron to Zanesville: How Are Ohio's Small and Mid-Sized Legacy Cities Faring?" *Greater Ohio Policy Center*. https://www.greaterohio.org/publications/akron-to-zanesville.

Hollingsworth, Torey, and Alison Goebel. 2017. *Revitalizing America's Smaller Legacy Cities: Strategies for Postindustrial Success from Gary to Lowell*. Cambridge, MA: Lincoln Institute of Land Policy.

Hope, Helen. 2021. "Announcing 2021 Driehaus Award Winner: The Evolution of Form-Based Codes." Smart Growth America. (October 12). https://smartgrowthamerica.org/2021-driehaus-winner.

Immergluck, Dan. 2017. "Sustainable for Whom? Large-Scale Sustainable Urban Development Projects and 'Environmental Gentrification.'" *Shelterforce*. (September 1). https://shelterforce.org/2017/09/01/sustainable-large-scale-sustainable-urban-development-projects-environmental-gentrification.

Interise. n.d. "Streetwise 'MBA': Our Award-Winning Program." https://interise.org/programs/streetwise-mba/.

International Economic Development Council. 2023. "Race, Equity, and Economic Development." https://www.iedconline.org/index.php?src=pages&ref=race-equity_preview.

Janus, Kathleen Kelly. 2022. "How Infrastructure Spending Can Help the Most Vulnerable Americans." *Stanford Social Innovation Review* (blog). (January 20). https://ssir.org/articles/entry/how_infrastructure_spending_can_help_the_most_vulnerable_americans.

Kannan, Vishnu, and Jacob Feldgoise. 2022. "After the CHIPS Act: The Limits of Reshoring and Next Steps for US Semiconductor Policy." Carnegie Endowment for International Peace. (November). https://carnegieendowment.org/2022/11/22/after-chips-act-limits-of-reshoring-and-next-steps-for-u.s.-semiconductor-policy-pub-88439.

Karner, Alex, Jonathan London, Dana Rowangould, and Kevin Manaugh. 2020. "From Transportation Equity to Transportation Justice: Within, Through, and Beyond the State." *Journal of Planning Literature*. (May). https://journals.sagepub.com/doi/abs/10.1177/0885412220927691.

Katz, Bruce, Peter Bassine, Della Clark, Gary Cunningham, Benjamin Della Rocca, Bulbul Gupta, et al. 2020. "Big Ideas for Small Business: A Five-Step Roadmap for Rebuilding the US Small-Business Sector, Reviving Entrepreneurship, and Closing the Racial Wealth Gap." Drexel University Nowak Metro Finance Lab. https://drexel.edu/~/media/Files/nowak-lab/201019BigIdeas_6P.ashx.

Kennedy. 2021. "Community Choice Aggregation Provides Leverage for Better Electricity Prices, Access to Solar." *pv Magazine USA*. (November 8). https://pv-magazine-usa.com/2021/11/08/community-choice-aggregation-provides-leverage-for-better-electricity-prices-access-to-solar/.

Khatana, Sameed. 2022. "The Increasing Death Toll in the US From Extreme Heat." *Time*. (July 20). https://time.com/6198720/heatwave-health-death-toll/.

Klein, Naomi. 2015. *This Changes Everything: Capitalism vs. The Climate*. Reprint edition. New York: Simon & Schuster.

Knight, Cameron. 2018. "The Time Cincinnati Tried to Protect the Poor from Pollution But Gave Up." Cincinnati.com/*Enquirer*. (December 13). https://www.cincinnati.com/story/news/2018/12/13/cincinnati-environmental-pollution/2289919002.

Kowalski, Kathiann. 2021. "Funding Challenges Limit Minority-Owned Businesses' Access to Energy Efficiency." *Energy News Network*. (January 4). https://energynews.us/2021/01/04/funding-challenges-limit-minority-owned-businesses-access-to-energy-efficiency/.

Leigh, Nancey Green, and Nathanael Hoelzel. 2012. "Smart Growth's Blind Side: Sustainable Cities Need Productive Urban Industrial Land." *Journal of the American Planning Association* 78(1): 87–103.

Lennon, Anastasia E. 2021. "Offshore Wind Staging Facility Coming to New Bedford Waterfront." *New Bedford Standard-Times*. (July 14). https://www.southcoasttoday.com/story/news/local/2021/07/14/offshore-wind-staging-facility-open-new-bedfords-waterfront/7962343002.

Lerner, Steve, and Phil Brown. 2012. *Sacrifice Zones: The Front Lines of Toxic Chemical Exposure in the United States*. Reprint edition. Cambridge, MA: MIT Press.

Lichten, Nathaniel, Joan Iverson Nassauer, Margaret Dewar, Natalie R. Sampson, and Noah J. Webster. 2017. "Green Infrastructure on Vacant Land: Achieving Social and Environmental Benefits in Legacy Cities." (February 15). *University of Michigan, Water Center* 34.

Lincoln Institute of Land Policy. 2022. "Comparative Cities Map." https://www.lincolninst.edu/research-data/data-toolkits/legacy-cities/comparative-cities-map.

Mallach, Alan. 2018a. *The Divided City : Poverty and Prosperity in Urban America*. Washington, DC: Island Press.

———. 2018b. *The Empty House Next Door*. Policy Focus Report. Cambridge, MA: Lincoln Institute of Land Policy. https://www.lincolninst.edu/publications/policy-focus-reports/empty-house-next-door.

———. 2022. *From State Capitols to City Halls*. Policy Focus Report. Cambridge, MA: Lincoln Institute of Land Policy. https://www.lincolninst.edu/publications/policy-focus-reports/state-capitols-city-halls.

Maryland Department of the Environment. 2022. "MD EJSCREEN." https://mde.maryland.gov/programs/Crossmedia/EnvironmentalJustice/Documents/mdejscreen-cejsc-2-25-2021v1.pdf.

Massachusetts Executive Office of Energy and Environmental Affairs. 2022. "Clean Energy and Climate Plan for 2050." (December). https://www.mass.gov/doc/2050-clean-energy-and-climate-plan.

MassDevelopment. n.d. "TDI Fellows." Transformative Development Initiative. https://www.massdevelopment.com/what-we-offer/key-initiatives/tdi/tdi-fellows.

MassINC. n.d. "Gateway Cities Innovation Institute." https://massinc.org/our-work/policy-center/gateway-cities.

Matthews, Tony, and Ruth Potts. 2018. "Planning for Climigration: A Framework for Effective Action." *Climatic Change* 148(4): 607–21. https://doi.org/10.1007/s10584-018-2205-3.

Mattiuzzi, Elizabeth, and Karen Chapple. 2020. "Epistemic Communities in Unlikely Regions: The Role of Multi-Level Governance in Fostering Regionalism." *Journal of Planning Education and Research*. (July 10). https://journals.sagepub.com/doi/abs/10.1177/0739456X20937287.

Mazella, Jon. 2021. "What Can State CIOs Do to Encourage Broadband Expansion?" *StateTech*. (August 24). https://statetechmagazine.com/article/2021/08/what-can-state-cios-do-encourage-broadband-expansion.

McFarland, Christiana, and J. Katie McConnell. 2013. "Small Business Growth During a Recession: Local Policy Implications." *Economic Development Quarterly* 27(2): 102–13. https://journals.sagepub.com/doi/10.1177/0891242412461174.

McGurty, Eileen. 2009. *Transforming Environmentalism: Warren County, PCBs, and the Origins of Environmental Justice*. New Brunswick, NJ: Rutgers University Press.

Morford, Stacy. 2023. "18 Huge, Billion-Dollar Disasters: Climate Change Helped Make 2022 the Third Most Expensive Year on Record. *Hoptown Chronicle*. (January 12). https://hoptownchronicle.org/18-huge-billion-dollar-disasters-climate-change-helped-make-2022-the-3rd-most-expensive-year-on-record.

Morley, David. 2014. "The Local Comprehensive Plan." *American Planning Association*. October 1. https://www.planning.org/publications/document/9007647.

Mulvihill, Keith. 2020. "Defender of Energy Efficiency—and Equity." NRDC. (February 3). https://www.nrdc.org/stories/defender-energy-efficiency-and-equity.

Nahm, Jonas, Joanna Lewis, and Bentley Allan. 2022. "Can the US Fight Climate Change—and Shift Industrial Policy?" *Washington Post*. https://www.washingtonpost.com/politics/2022/08/12/inflation-reduction-act-clean-energy.

National Association of Counties. 2010. *Growing a Green Local Economy: County Strategies for Economic, Workforce and Environmental Innovation.* (May). https://www.naco.org/sites/default/files/documents/Counties_Growing_Green_Local_Economy_-_June_2010.pdf.

National Complete Streets Coalition. n.d. "About the Coalition." Smart Growth America. https://smartgrowthamerica.org/program/national-complete-streets-coalition/.

National Consumer Law Center. 2016. "PACE Energy Efficiency Loans: Mortgages Still Risky Despite New Department of Energy Guidelines" National Consumer Law Center. (July 25). https://www.nclc.org/issues/pace-energy-efficiency-loans.html.

National Park Service. 2022. "New Bedford Whaling National Historical Park Visitor Center." Places. (February 25). https://www.nps.gov/nebe/planyourvisit/visitorcenter.htm.

New Jersey Department of Environmental Protection. 2022. "A Seat at the Table: Training for Whole-Community Climate Resilience Planning." https://coast.noaa.gov/digitalcoast/training/whole-community.html.

New Kensington Community Development Corporation. n.d. "Sustainable 19125 and 19134." https://nkcdc.org/community/cleaning-greening/sustainable-19125-and-19134.

New York State Climate Action Council. 2022. "Scoping Plan: Executive Summary." (December). https://climate.ny.gov/Resources/Scoping-Plan.

New York State. 2023. "Progress to Our Goals: New York's Nation-Leading Climate Agenda." New York State. https://climate.ny.gov/our-impact/our-progress.

New York State. n.d. Brownfield Redevelopment. Department of State. https://dos.ny.gov/brownfield-redevelopment.

NOAA National Centers for Environmental Information (NCEI). 2023. "Billion-Dollar Weather and Climate Disasters." https://www.ncei.noaa.gov/access/billions.

NY Renews. 2022. "Justice & Equity in the Climate and Community Protection Act." https://www.nyrenews.org/equity-memo.

Office of Sustainable Communities. 2010. "Partnership for Sustainable Communities: A Year of Progress for American Communities." US Environmental Protection Agency: Smart Growth. (October). https://archive.epa.gov/epa/sites/production/files/2014-06/documents/partnership_year1.pdf.

Organisation for Economic Co-operation and Development. 2006. "United States of America Workforce Investment Boards." https://www.oecd.org/cfe/leed/37728941.pdf.

PD&R Edge. 2022. "Pursuing Equitable Climate Adaptation in Legacy Cities." HUD User. (December 6). https://www.huduser.gov/portal/pdredge/pdr-edge-featd-article-120622.html.

Peduto, William, Jamael Tito Brown, Nan Whaley, Andrew Ginther, John Cranley, Steve Williams, Ron Dulaney Jr., and Greg Fischer. 2020. "Eight Mayors: We Need a Marshall Plan for Middle America." *Washington Post*. (November 22). https://www.washingtonpost.com/opinions/2020/11/22/marshall-plan-middle-america-eight-mayors.

Pelton, Tara, and Joseph Kane. 2021. "Weatherizing Homes Could Be One of the Most Vital Legacies of Biden's Infrastructure Plan." *Brookings* (blog). (April 22). https://www.brookings.edu/blog/the-avenue/2021/04/22/weatherizing-homes-could-be-one-of-the-most-vital-legacies-of-bidens-infrastructure-plan.

Perkins, Tom. 2019. "Publicly Owned Utilities 'Not a Panacea' but Can Produce Customer Benefits." *Energy News Network*. (December 16). https://energynews.us/2019/12/16/publicly-owned-utilities-not-a-panacea-but-can-produce-customer-benefits/.

Pew Research Center. 2023. "Economy Remains the Public's Top Policy Priority; COVID-19 Concerns Decline Again." (February 6). https://www.pewresearch.org/politics/2023/02/06/economy-remains-the-publics-top-policy-priority-covid-19-concerns-decline-again.

Philadelphia Water Department. n.d. "Green City Clean Waters." https://water.phila.gov/green-city.

Popkin, Matthew. 2022. "The Time is Ripe for Communities to Embrace Clean Energy on Brownfields." *RMI*. (September 26). https://rmi.org/time-for-communities-to-embrace-clean-energy-on-brownfields.

powell, john a., Stephen Menendian, and Wendy Ake. 2019. "Targeted Universalism." *Othering & Belonging Institute at U.C. Berkeley*. https://belonging.berkeley.edu/targeted-universalism.

Pulido, Laura, Ellen Kohl, and Nicole-Marie Cotton. 2016. "State Regulation and Environmental Justice: The Need for Strategy Reassessment." *Capitalism Nature Socialism* 27(2): 12–31. https://doi.org/10.1080/10455752.2016.1146782.

Purvis, Ben, Yong Mao, and Darren Robinson. 2019. "Three Pillars of Sustainability: In Search of Conceptual Origins." *Sustainability Science* 14(3): 681–95. https://doi.org/10.1007/s11625-018-0627-5.

Rajkovich, Nicholas, Terry Schwarz, Gwyneth Harris, and Adara Zullo. 2022. "Exploratory Scenario Planning for Climate In-Migration: A Guide for Cities in the Great Lakes Region." Working Paper. Cambridge, MA: Lincoln Institute of Land Policy. https://www.lincolninst.edu/publications/working-papers/exploratory-scenario-planning-climate-in-migration.

Reale, Hannah. 2021. "Local Governments Staff Up, Team Up to Confront Climate Change." WGBH. (December 8). https://www.wgbh.org/news/local-news/2021/12/07/local-governments-staff-up-team-up-to-confront-climate-change.

ReImagine Appalachia. 2021. "A New Deal That Works for Us." https://reimagineappalachia.org/wp-content/uploads/2021/03/ReImagineAppalachia_Blueprint_042021.pdf.

Resilient NJ. n.d. "Resilient NJ Local Planning for Climate Change Toolkit." Resilient NJ. https://experience.arcgis.com/experience/9daab51c2f5542969d50437522e012c4.

Reuters. 2021. "Solar Power Projects See the Light on Former Appalachian Coal Land." (December 30). https://www.reuters.com/markets/commodities/solar-power-projects-see-light-former-appalachian-coal-land-2021-12-30/.

Ricketts, Sam, Rita Cliffton, Lola Oduyeru, and Bill Holland. 2020. "States Are Laying a Road Map for Climate Leadership." *Center for American Progress* (blog). https://www.americanprogress.org/article/states-laying-road-map-climate-leadership.

Root, Tik. 2021. "Five Key Excerpts from the United Nations' Climate Change Report." *Washington Post*. (August 10). https://www.washingtonpost.com/climate-environment/2021/08/10/ipcc-report-un-takeaways.

Rosenbloom, Jonathan, and Chris Duerksen. 2022. "Saving the World Through Zoning." *Journal of Comparative Urban Law and Policy* 5(1): 363–374.

Rothstein, Richard. 2017. *The Color of Law: A Forgotten History of How Our Government Segregated America*. New York, NY: Liveright Publishing Corporation.

Rouse, David. 2022. "The Future of the Comprehensive Plan." *Journal of Comparative Urban Law and Policy* 5(1): 299–326. https://readingroom.law.gsu.edu/jculp/vol5/iss1/25.

Samarripas, Stefen, and Alexander Jarrah. 2021. "A New Lease on Energy: Guidance for Improving Rental Housing Efficiency at the Local Level." American Council for an Energy-Efficient Economy. https://www.aceee.org/research-report/u2102.

Schilling, Joseph, Kathryn Wertheimer Hexter, and Lauren Bulka. 2015. "Strong Cities, Strong Communities Fellowship Pilot Placement Program." The Promise of Urban Fellowships. (April). http://www.promiseofurbanfellows.com/wp-content/uploads/2016/01/Exec-Sum-HUDReport-4.1.15-.pdf.

Schilling, Joseph, and Gabriella Velasco. 2020. "Greenventory 2.0, Sustainability Lessons from Small and Midsize Legacy Cities." Working Paper. Cambridge, MA: Lincoln Institute of Land Policy. https://www.lincolninst.edu/publications/working-papers/greenventory-20.

Schneider, Keith. 2021. "Water Could Make Michigan a Climate Refuge. Are We Prepared?" *Circle of Blue* (blog). (February 16). https://www.circleofblue.org/2021/world/water-could-make-michigan-a-climate-refuge-are-we-prepared.

Schwartz, Nelson D. 2022. "Supply Chain Woes Prompt a New Push to Revive US Factories." *New York Times*. (January 5). https://www.nytimes.com/2022/01/05/business/economy/supply-chain-reshoring-us-manufacturing.html.

Shi, Linda. 2019. "Promise and Paradox of Metropolitan Regional Climate Adaptation." *Environmental Science & Policy* 92: 262–74. (February). https://doi.org/10.1016/j.envsci.2018.11.002.

Shuster, William, Stephen Dadio, Patrick Drohan, Russell Losco, and Jared Shaffer. 2014. "Residential Demolition and Its Impact on Vacant Lot Hydrology: Implications for the Management of Stormwater and Sewer System Overflows." *Landscape and Urban Planning* (May) 125: 48–56. https://doi.org/10.1016/j.landurbplan.2014.02.003.

Small Business and Entrepreneurship Council. 2021. "Facts & Data on Small Business and Entrepreneurship." https://sbecouncil.org/about-us/facts-and-data.

SolarYpsi. 2018. "About." May. https://www.solarypsi.org/about.

Sommer, Mark. 2017. "Mayor Signs Buffalo's Green Code into Law." *Buffalo News*. https://buffalonews.com/news/local/mayor-signs-buffalos-green-code-into-law/article_0ad80c32-064b-5a93-bfda-81633e95828e.html.

SouthCoast Today. 2009. "Mayor Robert Correia and Fellow Gateway City Mayors and Managers Unite to Promote Economic Development." (October 8). https://www.southcoasttoday.com/story/news/2009/10/08/mayor-robert-correia-fellow-gateway/51841282007.

Southeast Sustainability Directors Network. 2021. "Equity Resources." https://www.southeastsdn.org/programs/equity-resources/.

Stapleton, Jeremy. 2020. *How to Use Exploratory Scenario Planning (XSP)*. Policy Focus Report. Cambridge, MA: Lincoln Institute of Land Policy. https://www.lincolninst.edu/publications/policy-focus-reports/how-use-exploratory-scenario-planning-xsp.

State of California. n.d. "Strategic Growth Council. Transformative Climate Communities (TCC)." https://sgc.ca.gov/programs/tcc.

Steuteville, Robert. 2016. "Green Code Will Help Buffalo to Grow Again." Text. CNU. (December 28). https://www.cnu.org/publicsquare/2016/12/28/green-code-will-help-buffalo-grow-again.

Stutz. 2018. "With a Green Makeover, Philadelphia Is Tackling Its Stormwater Problem." *Yale E360*. https://e360.yale.edu/features/with-a-green-makeover-philadelphia-tackles-its-stormwater-problem.

Sustainable Development Code. n.d. "Sustainable Development Code Lists Land Use Provisions that Can Exacerbate Racial Bias." https://sustainablecitycode.org.

Sustainable Jersey. 2020. "Advancing Social Equity through the Sustainable Jersey Program: Analysis and Potential." Sustainability Institute at the College of New Jersey. (January). https://www.sustainablejersey.com/fileadmin/media/Grants_and_Resources/Publications/Equity_Report_Sustainable_Jersey_FINAL.pdf.

Tishman Environment and Design Center. 2019. *Local Policies for Environmental Justice: A National Scan*. (February). https://www.nrdc.org/sites/default/files/local-policies-environmental-justice-national-scan-tishman-201902.pdf.

Toxic Tours. 2022. "Southwestern Pennsylvania, United States." https://toxictours.org/us-pennsylvania.

TransForm. n.d. Fact Sheet on SB 375. https://www.ca-ilg.org/sites/main/files/file-attachments/TransForm_SB_375_Summary.pdf.

Tumber, Catherine. 2012. *Small, Gritty, and Green: The Promise of America's Smaller Industrial Cities in a Low-Carbon World*. Cambridge, MA: MIT Press.

Ungar, Lowell, and Steven Nadel. 2022. "Home Energy Upgrade Incentives: Programs in the Inflation Reduction Act and Other Recent Federal Laws." *American Council for an Energy-Efficient Economy (ACEEE)*. (September 28). https://www.aceee.org/policy-brief/2022/09/home-energy-upgrade-incentives-programs-inflation-reduction-act-and-other.

United States Conference of Mayors. 2019. "Mayors Climate Protection Center." United States Conference of Mayors. https://www.usmayors.org/programs/mayors-climate-protection-center.

US Bureau of Labor Statistics. 2013. "The BLS Green Jobs Definition." (January). https://www.bls.gov/green/green_definition.htm.

US Department of Agriculture. 2016. "Stewardship Mapping and Assessment Project: A Framework for Understanding Community-Based Environmental Stewardship." General Technical Report NRS-156. (January). https://www.fs.usda.gov/nrs/pubs/gtr/gtr_nrs156.pdf.

———. 2023. "Biden-Harris Administration Announces Historic Funding to Expand Access to Trees and Green Spaces in Disadvantaged Urban Communities." Press release. April 12. https://www.usda.gov/media/press-releases/2023/04/12/biden-harris-administration-announces-historic-funding-expand.

———. n.d. "Worcester's Urban Forest." Climate Hubs. https://www.climatehubs.usda.gov/hubs/northeast/project/worcesters-urban-forest.

———. n.d.a. "i-Tree." Climate Change Resource Center. https://www.fs.usda.gov/ccrc/tool/i-tree.

US Department of Energy. 2018. "Weatherization Works!" Fact sheet. (February). https://www.energy.gov/sites/prod/files/2018/03/f49/WAP-fact-sheet_final.pdf.

US Department of Housing and Urban Development. 2013. "National Disaster Resilience Competition." https://www.hud.gov/program_offices/economic_development/resilience/competition.

US Department of Transportation. 2023. "Federal Interagency Thriving Communities Network." (January 3). https://www.transportation.gov/federal-interagency-thriving-communities-network.

US Economic Development Administration. 2022. "Economic Recovery Corp." (November). https://www.eda.gov/funding/programs/economic-recovery-corps.

US Environmental Protection Agency. 1994. "Summary of Executive Order 12898—Federal Actions to Address Environmental Justice in Minority Populations and Low-Income Populations." https://www.epa.gov/laws-regulations/summary-executive-order-12898-federal-actions-address-environmental-justice.

———. 2008. "Reducing Urban Heat Islands: Compendium of Strategies." Draft. https://www.epa.gov/heat-islands/heat-island-compendium.

———. 2010a. "Partnership for Sustainable Communities: A Year of Progress for American Communities." Office of Sustainable Communities. (October). https://www.epa.gov/smartgrowth.

———. 2010b. "FY10 Brownfields Area-Wide Planning Pilot Program Award Recipients." https://www.epa.gov/sites/default/files/2015-09/documents/list_by_state.pdf.

———. 2013. "Codes that Support Smart Growth Development." Smart Growth. https://www.epa.gov/smartgrowth/codes-support-smart-growth-development.

———. 2015. "Distributed Generation of Electricity and Its Environmental Impacts." Overviews and Factsheets. (August 4). https://www.epa.gov/energy/distributed-generation-electricity-and-its-environmental-impacts.

———. 2019. "Brownfields Program Environmental and Economic Benefits." Overviews and Factsheets. https://www.epa.gov/brownfields/brownfields-program-environmental-and-economic-benefits.

———. 2021a. "Climate Smart Brownfields Manual." (June). https://www.epa.gov/land-revitalization/climate-smart-brownfields-manual.

———. 2021b. "EPA Announces $600,000 in Brownfields Technical Assistance Funding to Advance Equitable Development and Address Historical Environmental Justice Challenges in Brownfields Communities." News Release. (February 8). https://www.epa.gov/newsreleases/epa-announces-600000-brownfields-technical-assistance-funding-advance-equitable.

———. 2021c. "RE-Powering America's Land Initiative: Program Overview." https://www.epa.gov/sites/default/files/2021-03/documents/re_on_cl_program_overview_508_031121.pdf.

———. 2021d. "Revising Local Codes to Facilitate Low-Impact Development." LID Barrier Busters Fact Sheet Series. https://www.epa.gov/sites/default/files/2021-06/documents/lid_fact_sheet_codes_june_2021_508.pdf.

———. 2022a. "Codes That Support Smart Growth Development." Overviews and Factsheets. (December 23). https://www.epa.gov/smartgrowth/codes-support-smart-growth-development.

———. 2022b. "Integrating Green Infrastructure into Federal Regulatory Programs." Green Infrastructure. (July 25). https://www.epa.gov/green-infrastructure/integrating-green-infrastructure-federal-regulatory-programs.

———. 2022c. "Urban Runoff: Low Impact Development." Polluted Runoff: Nonpoint Source (NPS) Pollution. (July 25). https://www.epa.gov/nps/urban-runoff-low-impact-development.

———. 2022d. "Brownfields Job Training (JT) Grants." Brownfields. November. https://www.epa.gov/brownfields/brownfields-job-training-jt-grants.

———. 2023. "Bipartisan Infrastructure Law: A Historic Investment in Brownfields." https://www.epa.gov/brownfields/bipartisan-infrastructure-law-historic-investment-brownfields.

———. 2023a. "The Environmental Justice Thriving Communities Technical Assistance Centers Program." Environmental Justice. (April 27). https://www.epa.gov/environmentaljustice/environmental-justice-thriving-communities-technical-assistance-centers.

———. 2023b. "EJScreen: Environmental Justice Screening and Mapping Tool." EJScreen. https://www.epa.gov/ejscreen.

Vedachalam, Sridhar, Timothy Male, and Lynn Broaddus. 2020. "H$_2$Equity: Rebuilding a Fair System of Water Services for America." *Environmental Policy Innovation Center* 70.

Vibrant Cities Lab. n.d. "About/Contact." https://www.vibrantcitieslab.com/about.

———. n.d.a. "Step 8: Promoting Better Forestry on Private Lands." https://www.vibrantcitieslab.com/toolkit/promoting-better-forestry-on-private-lands/.

Vock, Daniel C. n.d. "Climate Migrants Are on the Move." *American Planning Association*. https://www.planning.org/planning/2021/winter/climate-migrants-are-on-the-move.

Washington Nature. 2019. "You Can Be a Voice for Healthy Urban Trees." The Nature Conservancy. (October). https://www.washingtonnature.org/fieldnotes/advocacy-urban-trees-puget-sound.

Wayne State University. 2023. "Detroit Revitalization Fellows." https://detroitfellows.wayne.edu.

Welter, Chris. 2021. "Why Is Utility-Scale Solar Coming to Ohio? Good Land and Plenty of Demand." WYSO. (August 15). https://www.wyso.org/local-and-statewide-news/2021-08-15/why-is-utility-scale-solar-coming-to-ohio-good-land-and-plenty-of-demand.

The White House. 2022. "Biden-Harris Administration Launches Version 1.0 of Climate and Economic Justice Screening Tool, Key Step in Implementing President Biden's Justice40 initiative. Press Release. (November 22). https://www.whitehouse.gov/ceq/news-updates/2022/11/22/biden-harris-administration-launches-version-1-0-of-climate-and-economic-justice-screening-tool-key-step-in-implementing-president-bidens-justice40-initiative.

The White House. 2023. "Building a Clean Energy Economy: A Guidebook to the Inflation Reduction Act's Investments in Clean Energy and Climate Action." CleanEnergy.gov. (January). https://www.whitehouse.gov/cleanenergy/inflation-reduction-act-guidebook/.

The White House. 2023a. "Fact Sheet: President Biden Signs Executive Order to Revitalize Our Nation's Commitment to Environmental Justice for All." https://www.whitehouse.gov/briefing-room/statements-releases/2023/04/21/fact-sheet-president-biden-signs-executive-order-to-revitalize-our-nations-commitment-to-environmental-justice-for-all/.

The White House. n.d. "Justice40 Initiative." https://www.whitehouse.gov/environmentaljustice/justice40.

Wilson, Bev. 2020. "Urban Heat Management and the Legacy of Redlining." *Journal of the American Planning Association* (May): 443–57.

Youngstown Neighborhood Development Corporation. 2020. Annual Report. https://www.yndc.org/news-media/yndc-publishes-2020-annual-report.

Acknowledgments

This report is dedicated to smaller legacy city communities and their national, state, and regional collaborators working on the ground to create innovative and equitable sustainability initiatives.

We could not have completed this work without an array of knowledgeable, steadfast readers, conversation partners, academic experts, and editorial professionals. Jessie Grogan of the Lincoln Institute of Land Policy supported this work from the start. She brought in Lincoln Institute colleagues as needed—notably Armando Carbonell, Maureen Clarke, Amy Cotter, Allison Ehrich Bernstein, Libertad Figuereo, Amy Finch, Katherine Gagen, and Daniel Janzow—and assembled a bank of three first-rate peer reviewers.

We are likewise indebted to Leah Bamberger, a former sustainability director for the City of Providence, and Alison Goebel, executive director of the Greater Ohio Policy Center, whose counsel much improved our work—and our thinking.

Special thanks also go to Reif Larsen, director of the Future of Small Cities Institute, who coproduced three webinars on greening smaller legacy cities in the lead-up to our report's release.

May this work inform and inspire smaller legacy city leaders seeking the new collaborations they will need to meet the urgency of climate resilience, equity, and green economic regeneration.

ABOUT THE AUTHORS

Joseph Schilling is a senior policy and research associate in the Research to Action Lab and Metropolitan Housing and Communities Policy Center at the Urban Institute. State and local governments serve as the primary platforms for his applied research, policy translation, and technical assistance work that helps cross-sector leaders adapt and transfer innovative policies and practices. Before coming to Urban, Schilling worked as a municipal attorney, a California legislative fellow, the director of community and economic development for the International City/County Management Association (ICMA), and a research professor of urban planning for Virginia Tech. Schilling's sustainability expertise includes research on HUD's Sustainable Communities Initiative and authoring a seminal American Planning Association article, "Greening the Rust Belt." While at Virginia Tech, he also led the initial design and development of the Eco-City Charter and Initiative of Alexandria, Virginia. In 2010, Schilling founded the Vacant Property Research Network, a hub for policy and research translation related to regenerating legacy cities.

Catherine Tumber is the author of *Small, Gritty, and Green: The Promise of America's Smaller Industrial Cities in a Low-Carbon World* (MIT Press, 2012). She is a Penn Institute for Urban Research scholar and a Gateway Cities Innovation Institute fellow with the Massachusetts Institute for a New Commonwealth. She holds a PhD and an MA in US history from the University of Rochester and a BA in social thought and political economy from the University of Massachusetts Amherst.

Gabi Velasco is a policy analyst in the Research to Action Lab at the Urban Institute, where their work focuses on environmental justice and housing justice. Previously, they worked with the sustainability program at the Texas Department of Parks and Wildlife, where they managed solar photovoltaic installations at Texas State Parks and conducted research on equitable greenspace access and sustainable architecture. Velasco received a BA in sustainability studies and a BA in urban political ecology, with a minor in women's and gender studies, from the University of Texas at Austin. While there, they also conducted community-engaged research, later published in the journal *GeoHumanities*, on environmental racism, zoning, and children's health in East Austin.